# Volume II of The Three-Fold Mimesis of Life

Mimesis Construction
Beginning to End

Where you start is not where you have to stay

## DR. RONALD BARNES

Copyright ©2025 by Dr. Ronald Barnes

All rights reserved. No parts of this book may be used or reproduced by any means, graphic, electronic, or mechanical, including photocopying, recording, taping, or by any information storage retrieval system, without the written permission of the publisher except in the case of brief quotations embodied in critical articles and reviews.

ISBN: 978-1-967375-94-3 (Paperback)
ISBN: 978-1-967375-95-0 (Hardback)
ISBN: 978-1-967375-96-7 (E-book)

Library of Congress Control Number: 2025921914

Printed in the United States of America

Published by:

info@thequippyquill.com
(302) 295-2278

# Table of Contents

Preface .................................................................................. 1

Introduction ......................................................................... 3

Chapter 1 **Pre-Natal Mimesis Concerns** ............................... 5

Chapter 2 **Stage 1 of the Three-Fold Mimesis – Birth to Nursery/Kindergarten** .......................................... 9

    General tips for parents in addressing the early years of a child's development: .......................... 9

Chapter 3 **Stage 2 of the Three-Fold Mimesis: Nursery through Elementary School** ............................. 17

    The Unique Needs and Strengths of Typical Young Children ................................................... 17

    Stage 2 of the Three-fold Mimesis – Secondary (High) School ...................................................... 21

    Family Structure ................................................ 24

    Children and Adolescent Future-Orientation Cognition Bias. ................................................... 28

    Factors Influencing Future-Oriented Cognitive Bias ................................................................... 29

    Why it is Important to socialize future orientation cognitive bias in kids ...................... 31

    Stage 2 of the Three-fold Mimesis– Post-Secondary School – College or University ...... 39

    Stage 2 of the Three-Fold Mimesis: Choices influence an individual's Life Experiences ...... 41

    Stage 2 of the Three-fold Mimesis: Job, Career, and Adult Responsibility ...................... 46

        Stage 2 of the Three-Fold Mimesis: Marriage
        Family – More Responsibility ............................ 54

        Stage 2 of the Three-Fold Mimesis: Becoming
        Parents .................................................................... 57

           The consequences of living in a dysfunctional
           family environment ........................................... 59

           Kids in two Parent-Homes vs Kids in Single
           Parent Homes ..................................................... 64

        Stage 2 of the Mimesis of Life ............................ 66

Chapter 4 **Stage 3 of the Three-Fold Mimesis of Life: The Twilight Years of Life: Reflective** .................... 71

        Stage 3 - The Final Mimesis of Life ................... 73

Chapter 5 **Social Psychology and Performative Interventions in Human Systems** .................... 77

Chapter 6 **Social Media Effect on the Child Developmental Process and on Society** ......... 81

        Beginning of Social Media ................................. 82
        The Emergence of Social Media ....................... 85
        Statistics on Gen Alpha Social Media Usage .. 87
        Is GEN Z and Gen Alpha Addicted to
        Technology? ......................................................... 88
        Social Media Usage among Youth and Teens . 94
        The Gloom and Doom of Social Media ........... 98
        Negative Consequences of Social Media ........ 99

Conclusion ........................................................................ 115

**End Notes** ..................................................................... 123

Volume II of the Three-Fold Mimesis of Life

# Preface

I am dedicating Volume II of the Three-Fold Mimesis of Life to my family. My entire family; Parents, Children, Brother, Nieces, Nephews, Grandparents, Aunts, Uncles, and Cousins. I am probably one of the rare exceptions in our society, then maybe not. I consider myself fortunate. Not one person in my immediate or extended family has ever experienced serious dysfunction. No one in my grandparents' generation, my parents' generation or my generation has ever had serious trouble with the law enforcement, legal system, and mental institutions or been to jail or prison; that I know of. That is a testament to both my fraternal and maternal Grandparents, and fortunately, I am the beneficiary of the Mimesis foundation they laid. The cycle of functionality has an inherent continuity characteristic, to be passed along to future generations, just as the cycle of dysfunctionality can be passed along to future generations.

    I want to especially dedicate the contents of Volume II – Chapters 1-5 to my parents. While writing this chapter and researching the Stages of the developmental process, I was constantly reflecting on how fortunate I am to have had grandparents and parents who were committed to family, each other, and their children, same as their parents before them. My parents laid the foundation for me to survive the challenges I have faced in life and proceed to construct a successful, positive, and happy Mimesis. Reflecting on my upbringing made me all the more insightful about how a stable and functional Stage 1 Mimesis is the foundation for Stage 2 Mimesis and life. Both of my parents are deceased. While my insight into Stage 3 Mimesis is based on research and the experience of my parents' passing, the most profound insight I have realized is an undeniable reality is that everyone dies. The influence of our Stage 1 Mimesis contributes to how we construct and experience Stage 2 Mimesis. My experience with Stage 1 and Stage 2 Mimesis leads me to believe, how we experience life is a foundation that influences how we will experience the end of life.

Volume II of the Three-Fold Mimesis of Life

# Introduction

Volume II of the Three-Fold Mimesis discusses the evolutionary details of Stages 1 thru three of Mimesis Construction. The various lived experiences that traditionally, individuals will encounter are addressed. The traditional lived experiences individuals encounter occur in multiple environments and stages of growth; the decision one makes results in Mimesis construction.

**Stage 1 Mimesis:**

The infant is in the home environment with parents or caregivers. They are the only influence the infant experiences for the most part.

**Stage 2 Mimesis:**

Nursery School, Elementary School, School, High School, Post High School, College, University, Work / Job environment, Life Experiences influenced by activities and individual's choices, What to do with your Life, embarking on a career or attaining a job, Thoughts about your future. Job. Career. Responsibility, Marriage, Family – More Responsibility, Becoming Parents; Stage 2 encompasses all.

**Stage 3 Mimesis:**

The natural evolution of living life leads into the twilight years. Reflecting back reveals the testament to how you have lived your life. How others perceive your life. The most challenging experience an individual will ever have in their lived experiences will be their experience, reconciliation and mitigation of dealing with your final Mimesis (death).

## CHAPTER 1

# Pre-Natal Mimesis Concerns

Some insight in understanding basics about the process of birth is important because according to medical science the cognitive process of a child becomes active before birth. Pre-natal at approximately 24 weeks the fetus becomes responsive to sound. The developing fetus is completely dependent on the mother for life. It is important that the mother takes good care of herself, diets healthy and receives prenatal care, which is medical care during pregnancy that monitors the health of both the mother and the fetus. Women, when pregnant, should avoid all teratogens. A teratogen is any environmental agent—biological, chemical, or physical—that causes damage to the developing embryo or fetus. Alcohol, smoking and drugs. Heroin, cocaine, methamphetamine, almost all prescription medicines, and most over-the-counter medications are considered teratogens.[1] If a fetus is exposed to teratogens, the odds of a child being born with deformities increase. Being born poor is a situation that can be overcome. Being born deformed is something that can last a lifetime.

Immediately upon the time, parents are aware they will have a child, their development role in the Mimesis construction of the child should become a forefront issue in their parental planning process. Being aware of the pre-natal development stages of a child will give parents time to plan and focus on the privilege and the responsibility of bringing a child into the world. The responsibility of parents in the Mimesis construction of the child begins when they become aware of the woman's pregnancy.

## Germinal Stage (Weeks 1-2)

A mother's and father's DNA is passed on to the child at the moment of conception. Genetics and DNA transference to the fetus. Conception occurs when the male sperm fertilizes the female egg and forms a zygote.[2]

## Embryonic Stage (Weeks 3-8)

After the zygote divides for about 7–10 days and has 150 cells, it travels down the fallopian tubes and implants itself in the lining of the uterus. Upon implantation, this multicellular organism is called an embryo. Now, blood vessels grow, forming the placenta. The placenta is a structure connected to the uterus that provides nourishment and oxygen from the mother to the developing embryo via the umbilical cord. Basic structures of the embryo start to develop into areas that will become the head, chest, and abdomen. During the embryonic stage, the heart begins to beat, and organs form and begin to function. The neural tube forms along the back of the embryo, developing into the spinal cord and brain.[3]

## Fetal Stage (Weeks 9-40)

When the organism is about nine weeks old, the embryo is called a fetus. At this stage, the fetus is about the size of a kidney bean and begins to take on the recognizable form of a human being as the "tail" begins to disappear.[4]

# Volume II of the Three-Fold Mimesis of Life

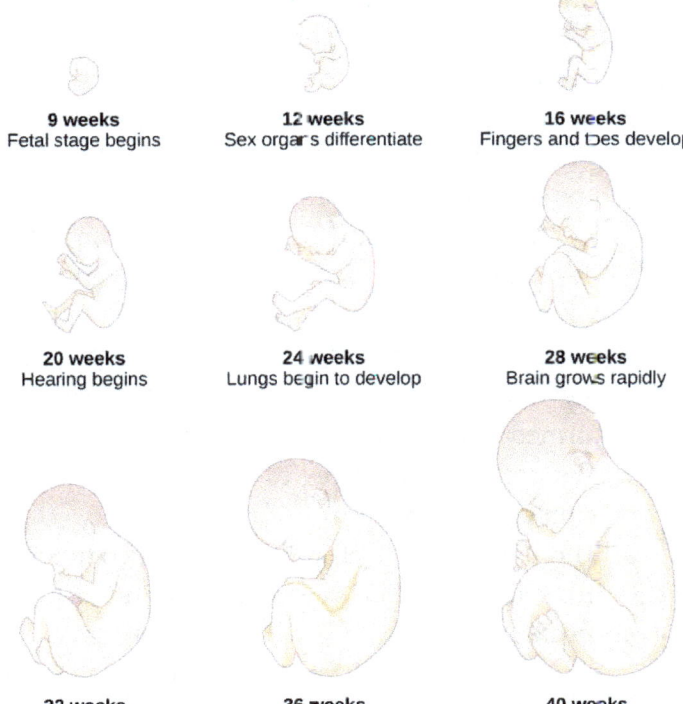

**9 weeks**
Fetal stage begins

**12 weeks**
Sex organs differentiate

**16 weeks**
Fingers and toes develop

**20 weeks**
Hearing begins

**24 weeks**
Lungs begin to develop

**28 weeks**
Brain grows rapidly

**32 weeks**
Bones fully develop

**36 weeks**
Muscles fully develop

**40 weeks**
Full-term development

- When the organism is about nine weeks old, the embryo is called a fetus. At this stage, the fetus is about the size of a kidney bean and begins to take on the recognizable form of a human being as the "tail" begins to disappear.
- From 9–12 weeks, the sex organs begin to differentiate.
- At about 16 weeks, the fetus is approximately 4.5 inches long. Fingers and toes are fully developed, and fingerprints are visible.
- By the time the fetus reaches the sixth month of development (24 weeks), it weighs up to 1.4 pounds. Hearing has developed, so the fetus can respond to sounds. The internal organs, such as the lungs, heart, stomach, and intestines, have formed enough that a fetus born prematurely at this point has a chance to survive outside of the mother's womb.
- Throughout the fetal stage, the brain continues to grow and develop, nearly doubling in size from weeks 16 to 28.
- Around 36 weeks, the fetus is almost ready for birth. It weighs about 6 pounds and is about 18.5 inches long,
- By week 37, all of the fetus's organ systems are developed enough that it could survive outside the mother's uterus without many of the risks associated with premature birth.

## CHAPTER 2

# Stage 1 of the Three-Fold Mimesis – Birth to Nursery/Kindergarten

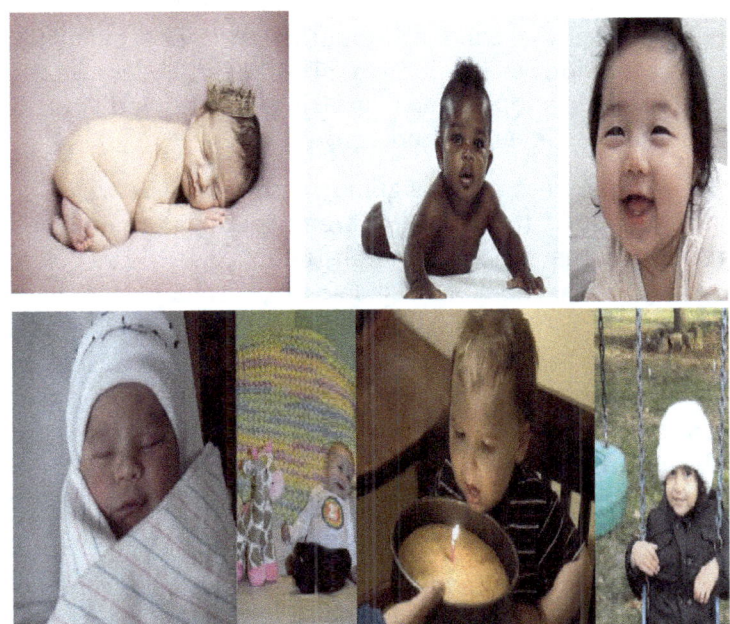

The Mimesis of life begins at birth. One might even be able to argue that the Mimesis of Life begins when the fetus is in the womb because research indicates that as the developing embryo becomes a fetus, it is affected by the mother's response to her lived experiences. The fetus develops instincts during gestation.[5] However, Stage 1 Mimesis formally begins when a child comes into the world with consciousness but without control or authority over the station or social location in which they are born.

There are many aspects of "being born" that are not within new born control, the culture socialization, the financial status, the place to live, nourishment, and guidance. A baby is born completely dependent on their parents or guardian for survival. Parents or guardians are the initial influence that gives guidance and establishes the initial perspectives the child will embrace. Mimesis being the representation of an individual's reality, is constantly evolving, developing and growing as the individual's experiences life evolves and grows.

One of the most significant processes by which an infant learns is thru social referencing. Social referencing refers "to the process wherein infants use the affective displays of an adult to regulate their behaviors toward environmental objects, persons, and situations. Social referencing represents one of the major mechanisms by which infants come to understand the world".[6] Often psychologist view social learning as a learned process of physical reactions and facial expressions a child learns and internalizes instinct to communicate with the adult caregiver.

Children who are exposed to adverse experience early in their developmental stage can demonstrate retarded cognitive and behavioral growth. Research on children experiencing high levels of neglect, and who fail to make proper infant attachments are at higher risk of experiencing slower or impaired brain development. A child who does not receive positive attachment from their primary caregiver can recover from this misfortune by experiencing attachments with another individual, considering *"The nature, timing, and duration of the experiences young children, launch them on a pathway of healthy development; also it is importance that intervention takes place early in the lives of children experiencing early neglect."* Considering the future success of society is dependent on the healthy development of its children, why American society has limitations in assuring the healthy development of all its children is a dilemma.[7]

Problems that arise from neglect at the early stages of development are poor impulse control, social withdrawal, problems with coping and regulating emotions, low self-esteem, pathological behaviors such as tics, tantrums, stealing and self-punishment, poor intellectual functioning and low academic achievement, and these are some of the few problems.[8]

Lack of positive attachment during the early stages of child development has significant and long-lasting impact, as does positive early childhood attachment. The lack of positive attachment can result in kids being challenged with higher-level functions such as cognition inhibitory control and working memory. Often neglected kids experience a delay in their ability to understand the mental states of others and it becomes challenging for them to regulate emotions. Neglected kids can suffer from high anxiety. One of the most common behaviors is indiscriminate friendliness. Which can lead to a child developing negative associations.[9] Stage one Mimesis ends and transitions to stage two Mimesis when the Child begins to have experiences outside of the home and when they have life experiences that are not monitored and influenced by parental or guardian oversight. For some kids stage one transitions to Stage 2 in early education, nursery school and for some children stage 2 begins at kindergarten.

## General tips for parents in addressing the early years of a child's development:[10]

- Be warm, loving, and responsive.
- Talk, read, and sing to your child.
- Establish routines and rituals.
- Encourage safe explorations and play.
- Make TV watching selective.
- Use discipline as an opportunity to teach.
- Recognize that each child is unique.
- Choose quality childcare and stay involved.
- Take care of yourself. How you care for yourself is an example for the child.

The issues today's youth deal with can become overwhelming: Climate Change, gun violence in the schools, drugs, social media, divorce, and virtual exposure to impersonal relationships/friendships. Kids today have a lot of "bad stuff" they might encounter in their lived experience, if parental or caregiver guidance is lacking. If feelings of hopelessness take hold of a kid's psychic and they fail to have hope for their future, kids will lack the ability to avoid anxiety and depression taking over their lived experiences.[11]

The high level of concurrent and sequential comorbidity (the simultaneous presence of two or more physical or mental diseases or medical conditions in a patient, i.e. anxiety and depression). Anxiety and depression in children and adolescents may result from (a) substantial overlap in both the symptoms and items used to assess these putatively different disorders, (b) common etiological factors (e.g., familial risk, negative affectivity, information processing biases,

Volume II of the Three-Fold Mimesis of Life

and neural substrates). Implicated in the development of each condition, and (c) negative previous anxiety condition causing increased risk for the development of depression. Basic research on their various common and unique contributing mechanisms has guided the development of efficacious treatments for anxiety and depressive disorders in youth. Potential processes through which the childhood anxiety is mitigated and depressive conditions are prevented, are first, a stable and loving family environment. Second, is counseling by child psychology professionals and as a last alternative, the child may need prescriptive drugs if their anxiety or depression becomes more serious and dangerous to their health or to the well-being of themselves or others.

Some children may demonstrate symptoms of anxiety, sadness and worry. A child may show signs of fear during their developmental process. Toddlers often show signs of distress and anxiety when separated (separation anxiety) from their parents, even if they are safe and cared for. Although some fears and worries are typical in children, persistent or extreme forms of fear and sadness could be due to anxiety or depression.

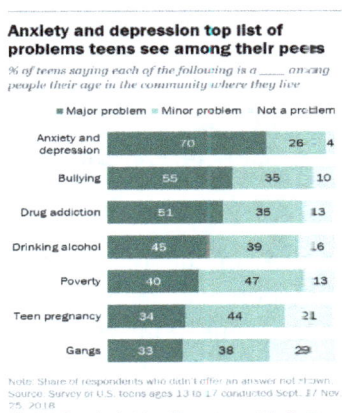

One has to wonder how teens can have anxiety and depression. The teenage years should be the most happy and fun fun-filled times of an individual's life. If teens are anxiety-ridden and depressed in their teenage years, then what will they become when they have adult responsibility?

## Anxiety and depression affect many children [12]

- 9.4% of children aged 3-17 years (approximately 5.8 million) had diagnosed anxiety in 2016-2019.

- 4.4% of children aged 3-17 years (approximately 2.7 million) were diagnosed with depression in 2016-2019.

## Anxiety and depression have increased over time [13]

- "Ever having been diagnosed with either anxiety or depression" among children aged 6-17 years increased from 5.4% in 2003 to 8% in 2007 and to 8.4% in 2011–2012.

- "Ever having been diagnosed with anxiety" among children aged 6-17 years increased from 5.5% in 2007 to 6.4% in 2011–2012.

- "Ever having been diagnosed with depression" among children aged 6-17 years did not change between 2007 (4.7%) and 2011–2012 (4.9%).[14]

Anxiety and depression are on the rise among America's youth, and whether they personally suffer from these conditions or not, seven-in-ten teens today see them as major problems among their peers. Concern about mental health cuts across gender, racial, and socioeconomic lines, with roughly equal shares of teens across demographic groups saying it is a significant issue in their community.

## Boys' and girls' goals and experiences differ in some key ways

*% of teens saying they ...*

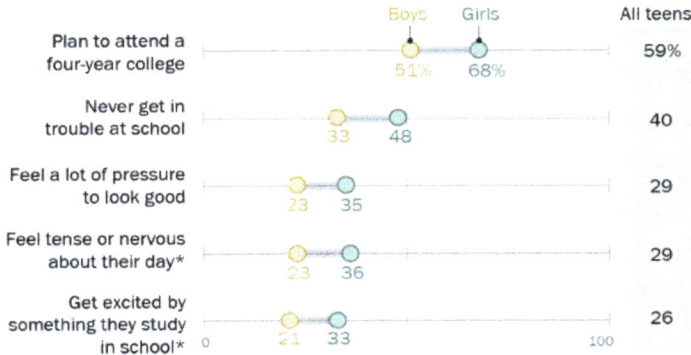

* Shares who say this happens every day or almost every day
Source: Survey of U.S. teens ages 13 to 17 conducted Sept. 17-Nov. 25, 2018.
"Most U.S. Teens See Anxiety and Depression as a Major Problem Among Their Peers"

**PEW RESEARCH CENTER**

Girls are more likely than boys to say they plan to attend a four-year college (68% vs. 51%, respectively), and they're also more likely to say they worry a lot about getting into the school of their choice (37% vs. 26%). Girls usually show more concern about their day-to-day lives and progress in school than boys do.[15] The concerns among today's youth indicate the need to have stability and love in their Stage 1 Mimesis. Stage 1 Mimesis stability will not eliminate the problems and challenges youth will encounter as they grow, but it may give kids a better ability to handle and deal with the challenges that will confront them as Stage 2 progresses.

Volume II of the Three-Fold Mimesis of Life

CHAPTER 3

# Stage 2 of the Three-Fold Mimesis

## Nursery through Elementary School

Nursery or elementary school is usually the time when children begin to have experiences outside of their parents' or guardians' control. At this stage, the child has more freedom to experience what their parental influence has taught them. Many of the child's decision-making and choices are not directly controlled by the parent on a consistent basis. In Stage 2, the child has their first experience with independence. They have some freedom away from parental oversight. The choices the child makes are influenced by their Stage 1 Mimesis, and the choices they make may be exploratory.

The choices they make can influence subsequent choices, some of which may follow them through their lives. As the Child develops, attends nursery or elementary school, accumulates friends and interact with other kids with whom they associate, their perspective about the world, life, and other people and especially about themselves, grows wider. They encounter more life experiences outside of the control of parents or guardians.

Their cognition learns to communicate, think, rationalize, and understand. Their personality is taking shape. These are the formative years. Characteristic of Mimesis Stage 2 is that the child, the individual, begins to accumulate life experiences to which they are catalyst. Kids at this stage begin to make decisions about friends and activities they become involved. Personality development continues forming during this stage as does the child's self-concept. Experiences are considered the engine that drives much of postnatal brain development. At this stage, kids absorb new learning like a succulent sponge that absorbs water. They live completely in the moment of the life experiences they are engaged.

Children ages 4 and 5 who have a positive Stage 1 experience are found to have characteristics considered important to succeed in preschool, such as curiosity, personal agency, level of dependency, and social competence. *"In the curiosity task (curiosity box) 84% of the most competent children had a history of secure attachment. The ones that performed the lowest were the ones with insecure-resistant histories of attachments. Children with histories of secure attachment also had higher scores in ego resilience measures. Securely attached children also faced less social problems in preschool and were able to handle problems in a more flexible way than children with histories of insecure attachments."*[16] According to childhood educators, children with a history of secure attachment were more socially effective, more positive, and less aggressive or fussy than the children with histories of insecure attachment.[17] The child's mind is like a sponge when it comes to learning. Every experience and encounter they have in their life is a learning experience. Children live in the now. They have no clear concept of the past or future. However, they do have a keen capability of retention for what they learn.

The following identifies the unique needs and strengths of typical young children, eleven important characteristics of primary learners (Ages birth – 8 years). Their ways of thinking and engaging with the world, and their remarkable hunger for learning. These characteristics are described by the English Learner (EL) Education network and are based on the research and writings of developmental psychologists and educators such as Lev Vygotsky, Maria Montessori, and Jean Piaget, in addition to peer-reviewed research and the experience of primary educators in the EL network. [18]

## The Unique Needs and Strengths of Typical Young Children

> Young children find security in rhythm, ritual, and repetition
>
> Young children learn through play
>
> Young children want to belong to a community that is safe, beautiful, and good.
>
> Young children explore the world with wonder.
>
> Young children "understand" the world first through their bodies
>
> Young children seek independence and mastery.
>
> Young children thrive in the natural world

Young children use stories to construct meaning

Young children seek patterns in the world around them

Young children construct their identities and build cultural bridges.

Young children express themselves in complex ways.

Parents or caregivers must pay attention and develop knowledge of their child's personality. The guidance and molding of a child's personality is largely the responsibility of the parent or caregiver. Given this, how a child develops is largely based on the influence of the parent or caregiver in the early stages of the child's life. Developmental Stages is a reference to the fact that the child is growing up and moving through the stages of evolution in their life. Children have an insatiable curiosity. According to the reference information on brain research from the California Department of Education, "birth to age three are the most important years in a child's development."[19]

## Stage 2 of the Three-fold Mimesis – Secondary (High) School

Stage 2 Mimesis is the longest period of Mimesis. It is a continuous cycle of individuals making life choices and accumulating experiences. The results of the choices they make continues through high school and last throughout the adult stage of life. The transition from elementary school to high school expands the individual's perspective. They become engaged in learning about more advanced concepts in education (science, mathematics, social studies, and other academic subjects), and in life, personally and socially. More of their self-concept and personality is exercised during this stage. The information and learning they experience regarding other people, their environment and themselves is also expanded. The life story of an individual is being written. This is the early chapters of the book of your life, (the early Mimesis construction process). The subsequent chapters are influenced by the chapters written up to this point in your life experience, (the Mimesis construction process).

The way a child behaves during elementary or the adolescent years, during the high school years of their development are a manifestation of their inner stability and security. *"All types of abuse are damaging to children—physically, emotionally, and psychologically—and can cause long-term difficulties with behavior and mental health development."* Child abuse might cause disordered psychological development and behavior problems.[20]

The World Health Organization defines Child Abuse as:

> *"[Child abuse is] all forms of physical and/or emotional ill-treatment, sexual abuse, neglect or negligent treatment or commercial or other exploitation, resulting in actual or potential harm to the child's health, survival, development or dignity in the context of a relationship of responsibility, trust or power.*[21]

> *"Child maltreatment is the abuse and neglect of people under 18 years of age. It includes all forms of physical and/or emotional ill-treatment, sexual abuse, neglect or negligent treatment or commercial or other exploitation, resulting in actual or potential harm to the child's health, survival, development or dignity in the context of a relationship of responsibility, trust or power. Four types of child maltreatment are generally recognized: physical abuse, sexual abuse, psychological (or emotional or mental) abuse, and neglect".*[22]

Volume II of the Three-Fold Mimesis of Life

## Global lifetime prevalence of Child maltreatment

WHO global status report on violence prevention 2014

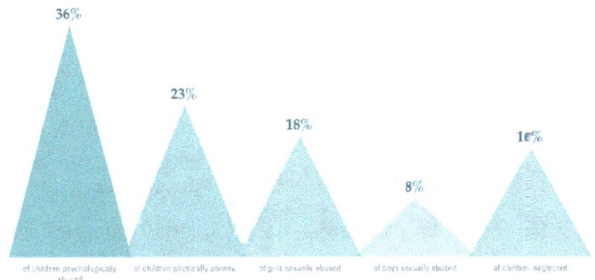

*36% of Children are psychologically abused*
*23% of Children are physically abused*
*18% of girls are sexually abused*
*8% of boys are sexually abused*
*16% of children are neglected*

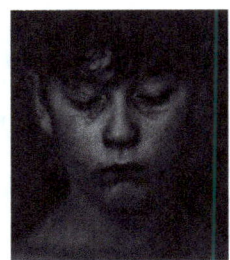

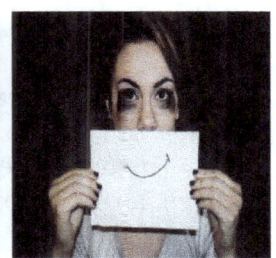

A child who can independently, without experiencing great amounts of anxiety, who easily establishes relationships with peers and adults, and is an active participant in his/her learning process, is considered a child who has received and is receiving a positive caregiver experience. Children are not always able to act independently, but they trust that while they explore the environment, someone will ensure that their safety is safeguarded, and their needs will be met.[23]

Children and adolescents who have had a positive attachment and experience a secure environment in the early stages of their developmental process are more likely to adapt positively to their environment in school and out of school.[24] Children with secure attachment histories were much more popular with peers and empathic than children with histories of insecure attachments.[25] The images above are consistent with kids who have developed a secure attachment with his/her caregiver(s). They are able to engage in positive, productive relationships with their peers.

## Family Structure

It is important to give some attention and focus to family structure because it has a significant impact on the development process of individuals, especially when the family demonstrates dysfunctional behaviors (abuse, divorce, neglect) Research indicates that family structure plays an important role in the health and well-being of children. Children living with their married, biological parents consistently have better physical, emotional, and academic well-being. One of the most devastating and impactful situations on an adolescent's life, is parental divorce. Research literature suggest that except for parents faced with unresolvable marital violence, children fare better when parents work at maintaining the marriage. [26] Attitudes toward marriage have changed over the years. Many young people think it is not necessary to be married to be together, or live together and have children without being married. Nevertheless, the important thing in a child's functional development is stability and functionality of the home environment. The child needs to experience his parents, guardians or caregivers constructing and maintaining a functional and positive home environment.

While there are many situations that can have a pivotal and traumatic impact on a child's or adolescent's life, divorce is among the most traumatic, except for the loss (death) of a parent or both parents. However, we will focus on divorce because it is the most common occurrence of trauma in a child or adolescent's life. From a Childs perspective divorce represents, a loss of family. When confronted with divorce children normally feel sad, angry, and anxious, and become confused to imagining how their lives with change. The age that a child experiences parental divorce and how the child will respond varies by the age of the child. If the child is introduced to a new family structure (remarriage), how that new family structure is constructed and how the child adjust is a variable factor to reestablish continuity, functionality, and stability back into the life of the child. [27]

The most common age of children when parents' divorce is between 7-14 years old. These are the years in a child's life when divorce has the most impact on their Mimesis development. Parent's divorce becomes a life experience that needs to be reconciled in their developmental process. Most first marriages that end in divorce, do so, between the years 7-8. Research evidence finds that approximately 50% of first marriages fail, around 60% of second marriages, and a staggering 73% of third marriages.[28] It is important for adults to take marriage seriously and consider the welfare of their children before and during their marriage.

## Age Guide on the Effects of Divorce on Children [29]

| Stage of Life | Effect of Divorce | Infant Reaction |
| --- | --- | --- |
| Infancy<br>Birth – 18 mos. | Babies can feel tension between parents but don't understand the reason. | -Irritability, clingy<br>-Emotional outburst<br>-May show signs of Slow Development. |
| Toddler<br>18 mos. – 3 years | Parents are the child's main bond.<br>Disruption to home is confusing.<br>Child feels responsible for breakups. | -Crying<br>-Need for attention<br>-Fear of abandonment<br>-Problem sleeping |
| Preschooler<br>3-6 years | Total Confusion / hard to comprehend.<br>Do not want parents to divorce.<br>Insecurity: Powerless to control outcome. | |
| School-Age Children<br>6-11 years | Fear of abandonment<br>Feelings of guilt/fault<br>Conscious fear of loss<br>Fantasize about reconciliation | |

Contemporary society values have changed over the years. Many young adults have a different concept of relationships than their parents or society tradition. The concept of being married has changed. People are waiting longer to marry and have children. This may be a good idea if they are more stable in their minds and behaviors about how to deal with and resolve life challenges, especially those that develop in a relationship This can, especially, be a good thing if by marring later in life and having children later in life makes adults better parents. There is a long-term negative affect divorce has on children. Research has found that many kids never fully recover from the impact of parental divorce on their lived experience.[30] The real problem is that interruption and negative life experiences can impact an individual for their entire life unless there is a successful reconciliation or intervention effort. Overcoming problems, difficulties, bad and challenging life experiences that happened in the earlier years lays a foundation for a cycle of such experiences, unless action is taken to break the cycle negative experiences has on the individuals. It is important to instill a cognitive awareness in the youth that helps to mitigate challenges the child will encounter in their lived experiences. A Cognitive Awareness bias will help reconcile, minimize or deter the youth from negative experiences. As the youth, progresses through Stage 2 Mimesis their future-orientation cognitive bias will increase as they go through life. As the emplotments in the narrative of life compile, the impact they have is contingent on how well the individual is prepared to respond to the life experiences they encounter.

## Children and Adolescent Future-Orientation Cognition Bias.

*"Future-oriented cognition bias is a cognitive bias describing the human tendency to favor positive future events over positive past events, and negative past events over negative future events. All else being equal, people prefer good things to lie ahead and bad things to be finished"*.[31]

Future-oriented cognition bias is a psychological concept of future orientation, which refers to an individual's focus on, anticipation of, and plans for the future. Adequate internalization of future-oriented cognition bias can serve as a cognitive deterrent for kids as they grow enabling them to make positive decisions in the variety of lived experiences they will inevitably encounter. Future-oriented cognitive bias has important implications for decision-making and well-being. Adolescents are instinctively prone to impulsive behaviors without giving adequate thought to consequences. It is normal that a youth will engage in some risk behavior. At the least, the youth should consider the results of their behaviors. "Future-oriented cognitive biases can lead to both good and bad decisions but a healthy future orientation bias can encourage planning and goal setting, while an unhealthy future orientation cognitive bias can lead to shortsighted or flawed choices".[32]

Generally, a child's age range is from birth to 12 years old. Adolescent age range is 10-19 years old. Generally, as long as kids depend on their parents or caregivers they are considered kids encompassing both, children and adolescents.[33] Over the age of 18 is generally considered *young adult.*

Adolescence:

13 to 19 years old (World Health Organization)

11 to 21 years old (American Academy of Pediatrics)

10 to 24 years old (some definitions)

It is important to note that these are general guidelines, and there may be variations in these age ranges depending on specific contexts and cultural definitions. However, there is a consensus that concludes that "children and adolescents do not universally exhibit a future-oriented cognitive bias; while young children may struggle with future thinking, late childhood and adolescence have a more complex indication. Adults typically display a future-oriented cognitive bias in "mind-wandering," but adolescents usually show no clear directional bias, or even a past-oriented cognitive bias. A future-oriented cognitive bias emerges with increased working memory, brain region development, and maturity. Future-oriented cognitive bias is important for goal setting and goal-directed behavior. Future-oriented cognitive bias development in children and adolescents is not consistent across studies and contexts".[34]

Young children, especially preschoolers, often display a present-focused perspective. Meaning their entire cognitive resource focuses on what they are experiencing in the present. Young children in the learning stages have not developed a forward-thinking mental apparatus. They display an inhibited future-oriented ability and an open-minded, uninhibited living in the present cognition. An open mind facilitates the learning process in children, adolescents, and kids. There is reason to believe that in late childhood (around 9-11 years old), kids have more future-oriented cognitive bias thoughts compared to past-oriented thoughts in mind-wandering tasks.

## Factors Influencing Future-Oriented Cognitive Bias

Typically, adults display a future-oriented cognitive bias in mind-wandering, but this pattern is not abundantly prevalent in adolescents, The presence and strength of a future-oriented cognitive bias can depend on the specific task, such as word cue tasks. A majority of past-event descriptions across all age groups, including adults stimulate future-oriented cognitive bias. For example, some kids might think, "I want to be a baseball player when I grow up. Other common forward-thinking biases are images of being a police officer, firefighter, a rapper, a basketball player. These images might be a result of the kid exposed to these images at a younger age. Boys play sports and envision sports figures as role models, likewise with authority figures, police, firefighters, doctors or lawyers. These future-oriented cognitive biases, however, are subject to change often during the course of the kid's development process into adulthood and as their lived experiences (emplotments) increase.

Working memory capacity, increases as kids age and experience more of life and as they become exposed to more of the environment. The cognitive development process of a kid from childhood into adolescence, into adulthood is linked to how their future-oriented cognitive bias is molded by the parent or caregiver. "Brain regions critical for future-oriented cognitive thinking, particularly the ventromedial prefrontal cortex, develop during childhood and adolescence, which can influence the emergence of this bias".³⁵ "The ventromedial prefrontal

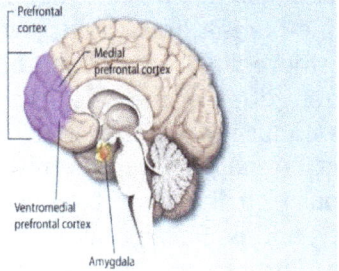

cortex (vmPFC) is a brain region crucial for decision-making, emotional regulation, memory representation, and the simulation of future scenarios".³⁶

Future-oriented thinking plays a key role in motivation and self-regulation, influencing long-term health, social relationships, and psychological well-being in adolescents. The formal learning process is a factor. As kids become exposed to subjects such as science, mathematics or social studies the may develop a proficiency. As their skills in learning increase, their motivation and self-regulation relating to future-oriented cognitive bias can become influenced. The same can be said for kids who participate in and are passionate about sports. They think about become a professional athlete. While a future-oriented cognitive bias is a key component of adult future thinking, its emergence and characteristics are complex and develop gradually across childhood and adolescence, influenced by socialization, environmental, cognitive and neurological factors. However a critical factor in developing future-oriented cognitive bias in youth is the influence and encouragement of the parent(s) or caregiver.

## Why It is Important to socialize future orientation cognitive bias in kids

The United States has the world's highest rate of children living in single-parent households where one parent is raising one or more children without the support of a spouse or partner, due to reasons like divorce, separation, death of a spouse, or a conscious choice to become a single parent. A single-parent family structure comes with numerous challenges, financial stress, increased responsibility, and a challenging work/family / personal time balance. Further complicating the single-parent family structure in America is that U.S. children are more likely to live in single-parent households without extended family support compared to children in other countries. The demographics compiled by the United States Census Bureau indicate that single-parent families are roughly 25-30% of U.S. children  In the U.S., 8% of children live with relatives such as aunts and grandparents, compared with 38% of children globally. In comparison, 3% of children in China, 4% of children in Nigeria, and 5% of children in India live in single-parent households. In neighboring Canada, the share is 15%. [37] Research studies indicate that children who grow up in single-parent homes *"face higher risks of experiencing emotional and behavioral problems, poorer academic outcomes, and challenges in social and economic development compared to those in two-parent homes. These difficulties are linked to factors such as parental stress, financial strain, and potential exposure to family instability or conflict, though not all children in single-parent families experience negative outcomes; many thrive with adequate support systems"*.[38]

Volume II of the Three-Fold Mimesis of Life

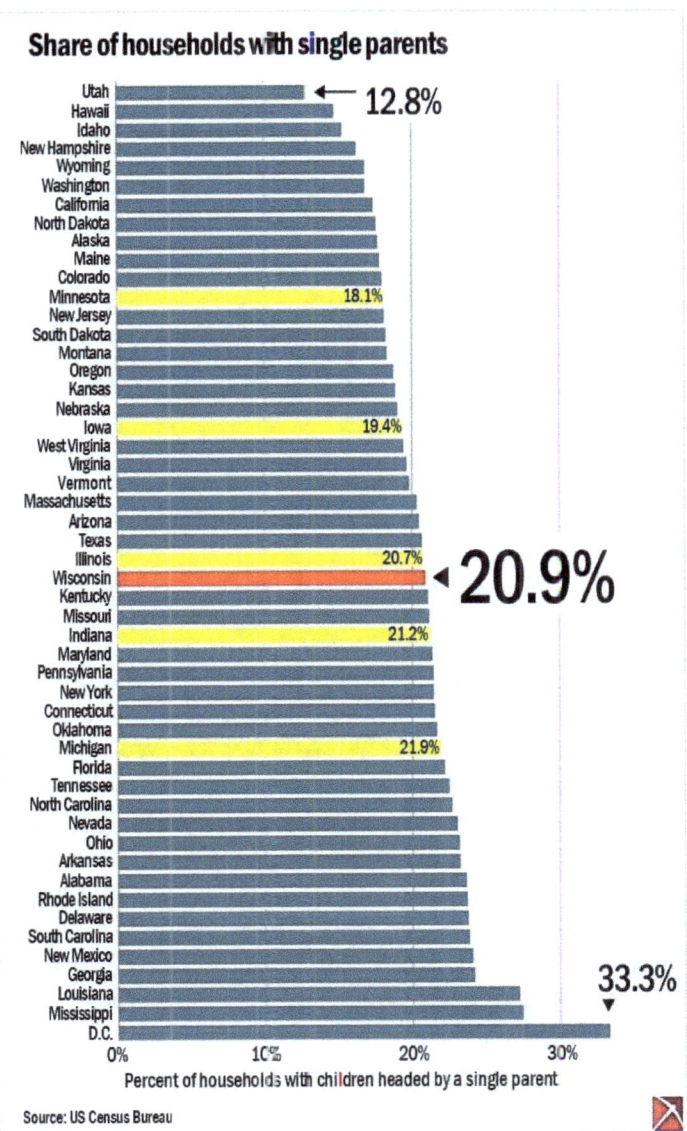

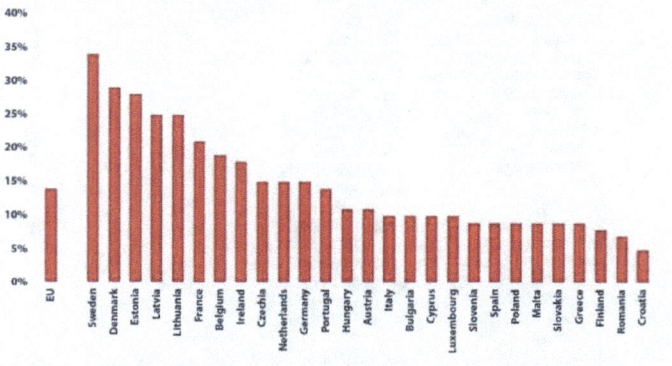

Further complication to single-family homes is the traumatic experience of children who lack constant adult supervision engaging in dysfunctional activity. If such dysfunction activity involves illegal activity and police intervention that creates a very significant problem for the single parent. Taking time off of work, the time and financial cost to intervene in the kids behalf are situations that make is all the more important to socialize the adolescent to make decisions that involve aforethought regarding the consequences of their actions. Twenty-five percent (25%) of all serious violent crimes involve a juvenile offender. Of these crimes, more than one-half involved a group of offenders. [39]

It is important for kids to understand they have agency over their Mimesis, the decisions they make determine the experiences they encounter. The single-family construct in America make it imperative for parents to instill a responsibility in kids to make decisions that guide their life on a positive trajectory, rather than engage troublesome situations. When kids have a future-orientation bias, they potentially consider the consequences of their actions.

Most youth stop engaging in lawbreaking as they mature and their brains develop. However, a lack of parental involvement can increase the risk of continued delinquency, the majority of young people "age out" of such behaviors. Supportive interventions, such as mentorship programs, job training, and educational support, can significantly reduce the likelihood of reoffending.[40] Crimes committed by juveniles have become more serious and have the potential to damage the life of the youth for many years to come in their future. The following information gives a brief profile of contemporary youth criminal engagement.

> *Violent crimes committed by White youth remained essentially unchanged during the study period, increasing by less than one-half of one percent (0.44%), while violent offenses committed by Black youth decreased by about 20%. Property crimes perpetrated by Black youth decreased by about 40%, while property crimes perpetrated by White youth decreased by roughly 52%.*[41]
>
> *Firearm involvement in juvenile offending was 21% higher in 2022 than in 2016, while other weapon use was 6% higher. This suggests a more pronounced increase in the use of guns relative to other weapons, rather than increased weapon use generally. Firearm use has also increasingly resulted in serious injury for victims in recent years.*[42]

In the 21ˢᵗ century, it is more imperative than ever to instill in kids a future orientation cognitive bias so they potentially take responsibility and acknowledge they have agency over their lived experience and Mimesis construction. This is something parents can teach that will give positive support to their parental responsibility.

A report from the U.S. Census Bureau concludes that 62% of new moms in their early 20s are unmarried. The report also found that 36% of all moms were unwed in 2011, up from 31% in 2005. In families with incomes of less than $10,000, that number goes up to 69%. [43]

Single moms are one of the most disadvantaged groups in the U.S.—nearly 30% of their families live under the poverty line, according to the US Census, as compared with 62% of families with married parents.[44]

- 18.4% of all births in the U.S. in 1980 were to unmarried women.
- 40.6% of all births in the U.S. in 2008 were to unmarried women.

In 1960, just 5 million children under 18 lived with only their mother. By 1980, that number had more than doubled. Today, according to the Annie E. Casey Foundation, 19 million children live in single-mother families, up from 17 million in 2000. In some school districts today, including several in New York and Michigan, a single mother leads the majority of families.[45]

Twenty-five million children are growing up without fathers in the home. That represents 40% of the kids in America. As reported by the Center for Children and Families:

- Forty percent (40%) of all live births in the US are to single mothers.
- Ninety percent (90%) of welfare recipients are single mothers.
- Seventy percent (70%) of gang members, high school dropouts, teen suicides, teen pregnancies, and teen substance abusers come from single-mother homes.

Statistically, a child in a single-parent household is far more likely to experience violence, commit suicide, continue a cycle of poverty, become drug dependent, commit a crime, or perform below his peers in education. [46]

- Sixty-three percent (63%) of suicides nationwide are individuals from single-parent families.
- Seventy-five percent (75%) of children in chemical dependency hospitals are from single-parent families.
- More than half of all youths incarcerated in the U.S. lived in one-parent families as children.

Thirty-seven percent (37%) of families led by single mothers nationwide live in poverty. Comparatively, only 6.8% of families with married parents live in poverty, according to data from 2009 compiled by the Heritage Foundation. Consider these dire statistics from single-parent households:[47]

- Sixty-three percent (63%) of youth suicides (Source: U.S. Department of Health and Human Services, Bureau of the Census).

- Ninety percent (90%) of all homeless and runaway children.
- Eighty-five percent (85%) of all children that exhibit behavioral disorders (Source: Center for Disease Control)
- Eighty percent (80%) of rapists motivated with displaced anger (Source: Criminal Justice & Behavior, Vol 14, p. 403-26, 1978.).
- Seventy-one percent (71%) of all high school dropouts (Source: National Principals Association Report on the State of High Schools.).
- Seventy-five percent (75%) of all adolescent patients in chemical abuse centers (Source: Rainbows for All God's Children.)
- Seventy percent (70%) of juveniles in state-operated institutions (Source: U.S. Dept. of Justice, Special Report, Sept 1988).
- Eighty-five percent (85%) of all youths sitting in prisons (Source: Fulton County Georgia jail populations, Texas Department of Corrections 1992)

The statistics on the plight of individuals who are in single-parent homes, according to statistics is dire. These statistics are from credible sources and describe a challenging situation for both the single parent and the kid raised in the single-parent home. Integrating the concepts of aforethought and future orientation cognitive bias into the socialization process of raising kids can potentially result in positive outcomes.

## Stage 2 of the Three-fold Mimesis– Post-Secondary School – College or University

Stage 2 of the Three-Fold Mimesis continues into the learning process in preparatory for the career path of an individual, the foundation for a job, career, and the future is laid. Some young adults go on to a higher education, college and some individuals do not go to college. They may go into a trade or enter the workforce directly out of secondary school. There is no prescribed path to find happiness or to be successful. However, society opportunities does favor those with a higher education. That does not mean an individual who has not been to college will not become successful or happy. Bill Gates, founder of Microsoft; Steve Jobs (deceased), founder of Apple computer; Richard Branson, founder of Virgin Records, Virgin Galactic Hotels and telecom companies; and Mark Zuckerberg, founder of Facebook aka Meta, are all college dropouts.[48] They are individuals who are extremely successful, in business. They made a lot of money and are among the richest people in the world. In American society, the amount of money one makes is considered successful. There are many who will argue the parameters that indicate success are much broader. Money happens to be only one and success in life is not defined by how much money one makes but also the quality of ones character, how they relate to others, how others relate to them, happiness, personal relationships, self-concept, the perception others have of you; are some of the other factors that feed the image of success.

The individual's life experiences are expanded in the after the "high school" stage of life. The Mimesis stage 2 continues. An individual's accumulation of life experiences is an ongoing process throughout Stage 2 Mimesis. Their life is being molded, shaped, and reinforced with each experience encountered. The emplotments of an individual's life narrative is adding chapters with each lived experience.

Volume II of the Three-Fold Mimesis of Life

## Stage 2 of the Three-Fold Mimesis: Choices influence an individual's Life Experiences

The experiences an individual encounters and becomes involved are catalyst that influence their Mimesis. Individuals will influence their future by the choices they make. How you think about phenomena you encounter, can influence how you act on it, and how you act on it can influence how you think, and the subsequent experiences you encounter. How an individual thinks, feels, and behaves influences their physiological well-being. For example, participation in sports, music, church, association with peers on a positive basis; all contribute positive to functional life experiences. Whereas participation in gangs, racist organizations and internalizing influences that have a negative influence, as determined by society, will result in likewise negative life experiences, such as incarceration or a premature termination of life. This is such an elementary concept, one would think it is common knowledge, and that alone will persuade individuals to make decisions that have positive influence, instead of decisions that will have negative influence. Ironically, that logic is not so definitive of all human behavior. Unfortunately, people make bad decisions, often, early in their life, which determines the experiences they encounter throughout their life. In a psychology research study on the topic of: *"Do children and adolescents have a future-oriented cognitive bias? A developmental study of spontaneous and cued past and future thinking."* The objective of the study was to determine if kids give thought to their future. Results found that more than 1/3 of younger children 6 to 15 years old, lacked foresight on the future and gave little evidence to produce future event descriptions.[49] In another study including 55% females and 45% Males, ages 22-33 years old, findings suggest that the view of the future is influenced by a temperamental and hereditary disposition.[50] This indicates the importance of an individual experiencing positive influence in their developmental process.

Individuals who had unresolved negative developmental experiences lacked future foresight. Research found that individuals 20 – 30 years old had the following three attitudes about their future:

- The first is *negative* future thinking (i.e., 'the future is uncertain and I can't do anything to modify it'), which is tied to specific temperament traits such as harm avoidance and reward dependence. [51]

> **Harm avoidance (HA),** a personality trait characterized by excessive worrying, pessimism, shyness, and being fearful, doubtful, and easily fatigued, is suggested to be related to low serotonergic activity.

> **Reward dependence (RD)** is characterized by a tendency toward dependence on signals of reward, especially verbal signals of social approval, social support and present mood state. As the results of other relevant research, results of our study show that the values of RD are correlated with depressive mood.

- The second attitude is characterized by *positive* future thinking (i.e., 'I believe in my abilities and that everything will be fine'). This attitude is negatively influenced by the harm avoidance trait. In other words, the lower the level of harm avoidance is, the more positive the perception of the future.[52]

- The third attitude is the a*voidance* of future thinking (i.e., 'I prefer not to think about future'). Although recent evidence has shown that avoidance—in both cognitive and behavioral processes—and depression are significantly correlated. The avoidant attitude seems to be completely independent from temperament. [53]

There are reasons, some people make good decisions and others make bad decisions. In many cases, the reasons are connected to the early development process of the individual as a child, toddler, and adolescent. However, it is not a firm rule that individuals who encounter early development depravation will have dysfunctional outcomes. Likewise, it is not a firm rule that individuals who experience positive and functional development experiences will have positive future experiences. However, the odds are higher a positive functional developmental process will result in a positive functional future and life.

There are studies that have found the ability to think and act on the thoughts reflecting future interest develops early in a child. The ability to develop and envision thinking about one's future is called *mental time travel:* Young children's ability to construct event sequences to achieve future goals is based on using details from the past to envision possible events, actions, and phenomena in the future. *"Future-directed thought is essential for complex cognitive activities, including planning for the future, accomplishing goals, and remembering to complete intended actions, that is, prospective memory"*. There is a saying, made in jest, by author George Bernard Shaw, "youth is wasted on the young".

Volume II of the Three-Fold Mimesis of Life

In Volume III, the lives of the following 15 well-known, high-profile individuals are reviewed. Evaluating their lived experiences will give clarity to concepts of how early lived experiences influence choices and circumstances later in life. The individuals profiled are:

| | |
|---|---|
| Dr. Martin Luther King, Jr. | Queen Elizabeth II |
| Muhammed Ali | Richard Nixon |
| Barak Obama | Malcom X |
| Oprah Winfrey | Stanley "Tookie" Williams |
| Billy Graham | Donald Trump |
| Jay-Z aka Shawn Carter | Tupac Shakur |
| Elvis Presley | Beyonce |
| Marilyn Monroe | |

One method parents or caregivers can use to develop *mental time travel* ability in a child is to expose them to a myriad of activities and experiences. This allows them to explore experiences and activities that capture their interest. Discovering a child's interest in the early stages gives them a focus upon which to develop and grow. While interest may change with subsequent experiences, the child will at least become engaged with experiences that give them choices.

# Stage 2 of the Three-fold Mimesis: Job, Career, and Adult Responsibility

Life is abundant with decisions, especially when individuals are at the threshold of making career choices. Not everyone is interested in college. Many satisfying careers do not require a college education. College education is a societal conceived prescription for success, theoretically. However, that is a concept that pigeonholes individuals into stereotypes for success. For many people college is not necessary for success or happiness. One aspect of the Mimesis construction process is for the individual who does not go to college to reconcile the deviation society unfairly places on the college vs no-college decision. The questions that confront individuals and the decisions they make are an indicator of the paths in life they will follow. Individuals who study medicine, science, law, creative arts or individuals who have a specific understanding of the career path they want, have already answered many of these questions. Most individuals are not so decided in their future and merely just want to find a job that pays them a good salary and satisfies their need for achievement and job satisfaction. Individuals who work in civil service, government work, construction or who are self-employed can have as fruitful a life as anyone else can. One of the shortcomings in our society is that people connect their happiness to their job and to money and material possessions. Practically, a positive orientation is to understand that happiness is one state of being and success is another state of being. Happiness is an internally sustained phenomenon that feeds life choices, including success, rather than allow life choices to feed whether or not one is happy or successful. The foundation and formula for happiness in life is implanted in the early stages of the developmental process.

For every step you take up the ladder of success there is someone who makes it possible.

    When an individual has job satisfaction, it is a reinforcing factor to their happiness. Life choices are more abundant when individuals experience and construct positive lived experiences, when the emplotments in their lived experiences align positively. As individuals advance in life, their Mimesis continues to accumulate life experiences. It is important to maintain a positive overall consciousness while advancing through life. Each experience in the emplotments of life are preparation for subsequent chapters in the Stage 2 Mimesis. Life's emplotments are accumulative, combined they narrate your life. How an individual aligns the emplotments in their life is important. For example, an individual can be a diligent employee during the day and a drug user / dealer in the evening. An individual can give the image of success that is only an illusion or snapshot into their reality. My father taught me a life-lesson: "If you can't stand up to it, don't do it".

Consider Bernie Madoff, the investment manager who constructed one of the largest Ponzi scheme investment frauds in United States history, cheating people who trusted him out of approximately $65 billion, many were scammed out of their life savings, under the guise of investing their money for retirement. He gave the appearance of an honest trustworthy individual. He was respected on Wall Street and had a good reputation among his clients, until his crimes were detected. An individual's Mimesis is based on the reality and accuracy of their life narrative. Some narratives of an individual's life do not tell the truth or only tell the partial truth. The entire truth eventually materializes. As the sayings go; "What's done in the dark will come to light" or "Everything comes out on wash day". Bernie Madoff is a classic example of the "tragic hero" (minus the hero). The Mimesis of Bernie Madoff is analogous to the character in a Shakespearian tragedy.[54] When the real estate crash of 2008 caused a tsunami repercussion throughout the financial industry, Clients of Bernie Madoff demanded withdrawals from their accounts he managed. The amounts clients requested were considerably more than the cash he had on hand or in any of the accounts associated with his business. Realizing his Ponzi scheme was crashing down around him, he confessed his crimes to his sons, Mark and Andrew, who both worked at the firm. After consulting with a lawyer, Mark and Andrew turned their father in to the federal police, who arrested him at his New York City penthouse apartment the following day. Despite their employment at the firm, they repeatedly denied any involvement in their father's Ponzi scheme. [55] The disgrace and betrayal of his father was too much for Mark to bear. In 2010, Mark committed suicide. In 2014, Bernie Madoff's other son Andrew, died of cancer. [56] This is a real life version and representation of a Shakespearian tragedy. It is also an example of how a negative Mimesis construction results. An individual can be respected at one moment in their life, and respond in a negative way to phenomena that confronts them then discover that is the first step toward their demise. It takes 1,000 deeds to build trust and only one negative deed to destroy the trust that was built.

    Securities Exchange Commission (SEC) Agents, Harry Markopolos and Frank Casey, were alerted, given that Madoff's supposed trades should have had a substantial ripple effect on broader markets, Markopolos suspected that Madoff was not trading. With the assistance of two of their colleagues, Neil Chelo and Dave Fraley, Markopolos and Casey continued to probe the Madoff operation. Their findings supported the claims of Mark and Andrew Madoff that Bernie Madoff conducted fraudulent crimes.[57]

    Often you will hear people say, "I pulled myself up by my boot straps", or "I made it on my own", or "I am a self-made person". All of these statements are at the least misconceptions; most are exaggerations. Many are self-aggrandizing arrogance. In reality, they are not true. For every door of opportunity that opens, someone else has the key that opens the door for the individual who enters. No one makes it on their own. Everyone who makes it has a helping hand or helping hands from others somewhere along the way or most often at every step of the way. People can also close the door on you, denying you entry to opportunity.

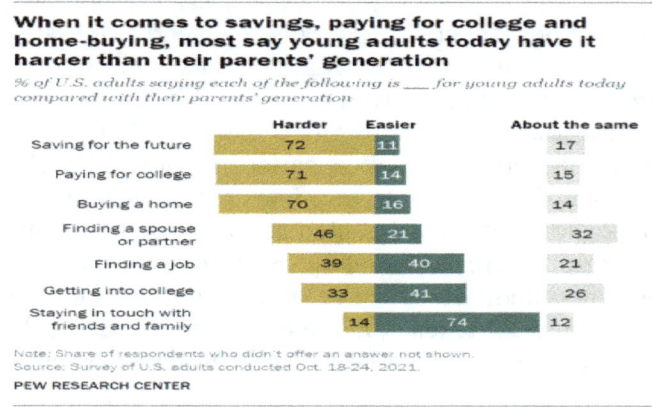

The graph above represents results from a study conducted by the Pew Research Center that examined various areas of life from the perspectives of young adults, identifying the challenges they face in comparison to those faced by their parent's generation. The results of this study indicate the importance of parents creating a stable and positive developmental process for their kids and for kids to create and reinforce, for themselves, positive lived experiences that align.[58] It also speaks to the foundation of society and the concern about the quality leadership that governs America. Most people who govern America, from the President of the United States on down, give more attention and advantage to those with money than they do to the masses of American citizens. In other words, for decades, America leadership has been more responsible to the 1% rich who control over 30% of the nation's wealth, and who are a smaller percentage of the population. Fewer numbers of people are considered "rich" compared to a larger segment that identifies as middle or lower class, yet the minority controls the destiny of the majority. That being the case, the facts are that their interest in the minority has always taken a backseat to the interest of the rich minority. Frankly, the rich do not care about America. They care about their money; first and foremost. Recent data from 2023-2024 shows the top 1% of households holds over $38 trillion in wealth, far more than the combined wealth of the middle class. This is not a condemnation of the wealthy. The voters who keep electing partisan leadership who neglect their interest are at fault. However, many of the wealthy rely on the masses to become wealthy. They sell their products and services to the masses and they secure government grants, loans and subsidies that come from taxpayer dollars.[59] The construct of American economics needs revision for the best interest of America and the American economy.

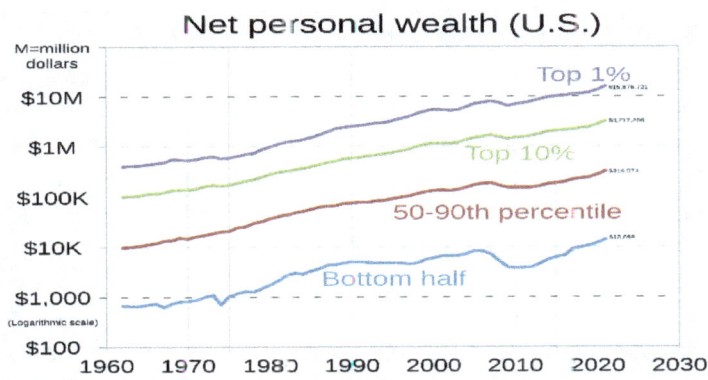

## Stage 2 of the Three-Fold Mimesis: Marriage Family – More Responsibility

It is a traditional societal value and a positive life experience for individuals to marry and have a family. A good marriage is a stabilizing factor in an individual's life, for those who commit to the marriage. A good marriage is an advantage and positive factor in the way others view the individual. For individuals to be on a stable and positive life course when they bring another person into their life indicates a positive Mimesis construction. Planning a marriage and family, in general, results in a better life experience than having babies without proper planning or as a single parent (a generalized traditional opinion). While most births (59%) are to married parents, statistics show that 90% of teens who get married, do so because of pregnancy and will be divorced within six years. Unless individuals are 100% sure that they are in love, and that they want to be married to each other for the rest of their life, then they should consider other options for the sake of each other and especially, for the sake of the baby, if pregnancy factors into their reason to marry.[60] This section on the Stage 2 of Mimesis life experiences involves one of the most important decisions an individual will make in their life, the taking of a spouse, a life partner, and creating a family. The wrong choice can bring havoc into your life. The right choice can bring immense happiness, harmony and joy, leading to a fulfilling life. The previous stages of Mimesis and life experiences an individual encounters is one determinate factor in choosing a spouse and how compatibly you align.

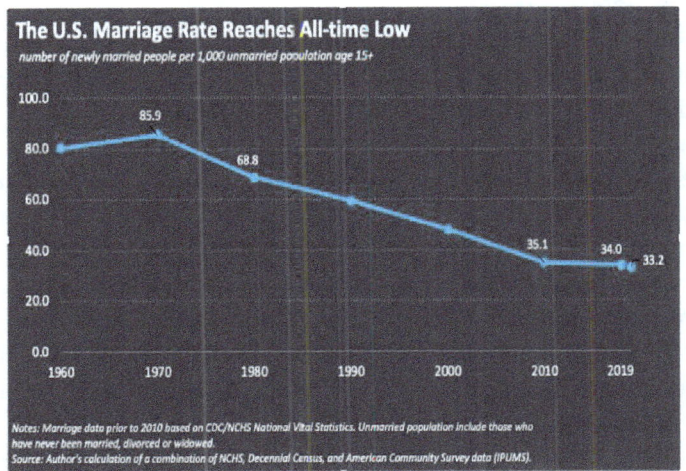

In America, surprisingly, the divorce rate is decreasing. A lower divorce rate means longer marriages.[61] The marriage rate in America is also at an all-time low because people are waiting longer to get married.[62]

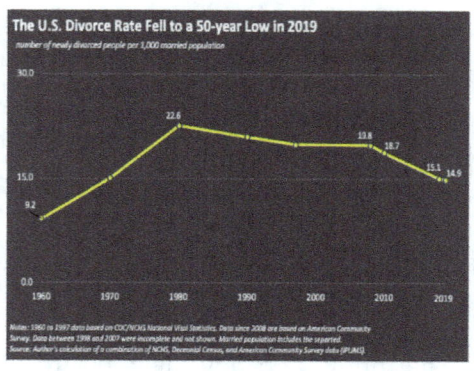

In 1914, the divorce rate in America was 1%. In 1981, the divorce rate hit an all-time high in the United States of 53% of all marriages ending in divorce. It is estimated that the divorce rate for 2021 will be 45%.[63] One can conclude that because individuals are waiting longer to marry, they are more mature and committed, thereby better prepared for marriage.

Divorce shatters dreams, crushes hopes, breaks wedding vows, confuses, depresses and saddens children, damages the children and both spouses in one way or another. Divorce creates a different emplotment of experiences that challenge the Mimesis in life. On the other hand, the opposite of divorce is a long-lasting marriage and a happy family that gives a positive emplotment to the Mimesis in life. Marriage is challenging but also rewarding. It takes mature responsible people to maintain a long-lasting, happy marriage. A healthy marriage has a significant impact on Mimesis and life experiences in very positive ways. There is a rewarding fulfillment in contributing to the happiness of others. Few things are more satisfying to a parent than to witness their kids be happy and grow into fulfilling satisfying lives. When parents create a positive happy environment reinforced with love for their children to thrive and grow in, that gives a positive foundation for the Mimesis construction of the youth.

## Stage 2 of the Three-Fold Mimesis: Becoming Parents

If there is any truth to the term "Poetic Justice", it is realized and understood when individuals become parents themselves. Upon become parents, individuals get a taste of their own medicine. They can gain a full understanding of what their parents endured. One can argue that the most important job in society is having and raising kids, being a parent. The responsibility for shaping the life of another individual is a monumental task. It can be very rewarding or very disappointing. Parenting, unfortunately, is not taught in school. There is no formal training for being responsible for an individual and shaping their life from birth to the standard approximate age of 18 years old. Becoming a parent involves the process of raising a child from birth to independent adulthood. Facilitating the upbringing of a child through stage 1 and the foundational emplotments of stage 2. Caring for and nurturing a child, fulfilling the parental responsibilities that accompany child raising is often influenced by the previous lived experiences of the parents. As a parent, foremost in your mind and relationship with the child should be the ever-present thought that you have responsibility for the way their life turns out. The parent or guardian is an agent/catalyst to the plight in life that befalls the individual they raise. Unfortunately, in today's society parents range in age from teenagers to mature adults. Teenage parents are comparable to "babies raising babies". The parent you are, and were, is revisited in the reflection stage of your Mimesis. As your adult lived experiences accumulate, the outcome of the individual you were responsible for raising will become a factor in your own self-evaluation. The Mimesis of your child may be reflective of the lived experiences of the parent or caregiver. Lived experiences of the parent, often have a cycle of continuity cycle that is reflected in the child.

Understanding the differences and characteristics of a functional family environment is important. Often individuals who experienced a dysfunction family environment believe dysfunction is normal.[64]

| Characteristics of a functional family Environment | Characteristics of a nonfunctional/dysfunctional family Environment |
|---|---|
| Several characteristics are generally identified with a well-functioning family. Some include: support; love and caring for other family members; providing security and a sense of belonging; open communication; making each person within the family feel important, valued, respected and esteemed. | A dysfunctional family **is** a family in which conflict, misbehavior, and often child neglect or abuse on the part of individual parents occur continuously and regularly, leading other members to accommodate such actions. Children sometimes grow up in such families with the understanding that such a situation is normal. |

Volume II of the Three-Fold Mimesis of Life

## The consequences of living in a dysfunctional family environment

Early Childhood exposure to a dysfunctional family environment can cause problems and injuries that follow an individual through their entire life. The potential to suffer from psychological disorders such as depression, anxiety, and addiction increased with dysfunction. Children are vulnerable to developing psychological or behavioral disorders.

The impact of dysfunctional early childhood exposures can cause the following behaviors in kids and manifest in their adult year.[65]

1. **Rebel:** not only rebel against the authority of the parents, but have problems with all those who have some power, from teachers to the police. These children often end up being labeled as "problematic" and develop behavioral problems.

2. **Scapegoat:** this is a child who has been accused of most family problems, so he developed a deep sense of guilt that can turn him into the boxing sack of others by adult.

3. **Guardian:** This child usually takes on the role of parents, so it grows too fast and loses much of his childhood because he has solved family problems alone or has been a mediator in adult conflicts.

4. **Lost:** he is a discreet, quiet and timid child, whose needs were ignored, so he learned to hide and repress his emotions. He usually becomes an adult who believes he is not worthy of being loved because he does not have a good self-esteem.

5. **Manipulator:** is an opportunistic child who exploits the mistakes and weaknesses of other members of the family to get what he wants.

It is an unfortunate situation for a child to experience dysfunction in their early developmental process. One has to wonder if adults, parents, caregivers or guardians are aware of the impact they have on a child's life. Some adults are only reflecting their own experience when they become parents or caregivers themselves. Breaking the cycle of dysfunction is important for the child and the parent. It can be a benefit to parents, who themselves have experienced childhood trauma or traumatic life experiences, to undergo counseling before becoming parents.

Being a parent is much more than having a baby. It is, arguably, the most important job an individual will ever have. It is a 24-hour/7-days-a-week/and 365-day-a-year responsibility,[66] from the time of conception (pregnancy) until the child reaches adult independence. The most important job in society is shaping young minds, preparing individuals to become the future generation of leaders who direct society and who themselves will become parents. Ideally, two parents, traditionally a man and woman, should raise a child. However, family structures have evolved such that people of the same sex in a relationship can raise kids, and more commonly, today, is that single parents raise many kids. Regardless of the family structure in which a child is raised and socialized, the most important factor is for the child to be exposed to a functional, loving, and secure home environment.

According to Olsen's Circumplex Model, there are five parenting styles **Balanced, Uninvolved, Permissive, Strict, and Overbearing**.[67]

> **The Balanced style** "is considered optimal because there is a balance of separateness versus togetherness on cohesion and a balance of stability versus change on flexibility. Balanced style parenting is moderate to high on both closeness and flexibility. The area of Balanced parenting style is larger than any one of the other four styles, acknowledging the diversity of positive ways parents can raise children well. The Balanced parenting style is characterized by warm and nurturing parents who are supportive emotionally, responsive to their child(ren)'s needs, encouraging toward independence (with monitoring), consistent and fair in meeting out discipline, and who expect age-appropriate behavior."

> **The Uninvolved parenting style** "is very low in closeness between parents and child(ren) and very high in flexibility. The Uninvolved parenting style is characterized by low emotional connection, low responsiveness from parent to child, high independence of child from parent (parents are disconnected from child's life), highly negotiable rules that are loosely enforced, and few demands made on the child."

> **The Permissive parenting style** "is very high in closeness between parents and child(ren) and also very high in flexibility. The Permissive parenting style is characterized by parents who are overly protective of their child(ren), very responsive to their child(ren)'s every need, more of a friend to their child(ren), lenient in discipline, and unlikely to place demands on their child(ren)."

> **The Strict parenting style** "is very low in closeness between parents and child(ren) and also very low in flexibility. The Strict parenting style is characterized by strictly enforced rules, highly restricted child freedom, firm discipline, low responsiveness to child, and low emotional connection between parent and child."

> **The Overbearing parenting style** "is very high in closeness between parents and child(ren) and very high in flexibility. The Overbearing parenting style is characterized by overly protective parents who cater to the child's every need and act more like a friend to the child while at the same time strictly enforcing a proliferation of rules with firm discipline."

The Olson Circumplex Model found that approximately 33% of the couples studied responded as having the same parenting style. The **Balanced Style** (a balance of cohesion and flexibility in parenting) is considered the healthiest. The **Permissive parenting style** had the most negative outcomes, with the **Uninvolved style** reporting outcomes just as negative. Most parents reported having similar types of problems with their children regardless of their parenting style. The resulting difference in the kids is determined by how lived situations are handled by the parents. The parenting styles determine how parents handle the life situations they encounter with their kids.

Parenting affects the relationship between the parents. There is now a third person to be cared for, a third person who is dependent and will require considerable time and attention.

There is much concern about whether having a baby causes substantial declines in the average couple's relationship. This is an important concern because the first child is often born within the first five years of marriage. This period has the highest risk for divorce.[68] The quality of a couple's relationship following a baby's birth has critical implications for numerous aspects of the baby's early development, involving their mental (psychological), physiological attachment, and language development. Late or retarded child development (e.g., psychological, social, and school functioning) is also related to the status and quality of parent relationships.[69] While single parenting does put a heavier burden on the parental individual or caregiver, giving a child the proper care does increase the potential for positive development of the child. Depriving the child will, conversely increase the potential for the child to develop dysfunctional characteristics. The maturity factor in raising a child cannot be overstated. A teenage pregnancy causes extremely difficult problems for the child, the young teenage parents, and the parents of the teenager. Good parenting will not necessarily prevent a teenage pregnancy, but it will reduce the odds of it happening. A young person with the right focus in their life will embrace lived experiences that align with their upbringing and objectives in life.

Dr. Ronald Barnes

# Kids in two Parent-Homes vs Kids in single Parent Homes

In 2019, only 63% of children in the U.S. lived with married parents, down from 77% in 1980. This decline has not been experienced equally across the population. There has been little change, for example, in the family structure of children whose mothers have a four-year college degree: In 2019, 84% of children whose mothers had four years of college lived with married parents, a decline of only 6 percentage points since 1980. Meanwhile, only 60% of children whose mothers had a high school degree or some college lived with married parents, a 23% point drop since 1980. A similarly large decline occurred among children of mothers who did not finish high school; the share of children in this group living with married parents fell from 80% in 1980 to 57% in 2019. The graph below plots these trends. [70]

**Percent of children living in married-parent households by maternal education**

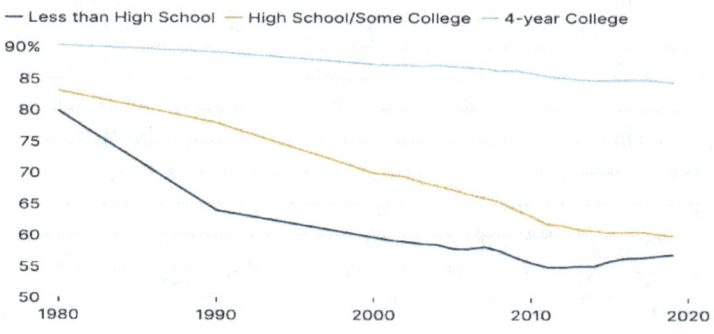

Source: Author's calculations using 1980 and 1990 US Decennial Census and 2000-2019 US Census American Community Survey.
Note: Observations are weighted using child's survey weight.

BROOKINGS

> "Marriage rates have fallen significantly in the U.S., including among adults who have children. It is important to recognize that these changes in marriage patterns are not without consequence for raising children. Data show conclusively that parents are not cohabitating without marriage in a way that remotely accounts for the decrease in two-parent married households. These kids are more often living with one parent. Data shows that these trends are highly correlated with the parents' education: The share of children living in two-parent homes is 71% when the mother has only a high school degree and 70% when she does not have a high school degree. A much higher share, 88%, of children of mothers with a four-year college degree live in a two-parent home."

Children who grow up with one parent in the home are at a considerable disadvantage relative to kids who grow up with two parents in the home. This does not mean that children raised by a single mother cannot be successful in life. Many are. Considerable research indicates that the odds of graduating high school, getting a college degree, and having high earnings in adulthood are substantially lower for children who grow up in a single-mother home. The odds of becoming a single parent are also substantially higher for children who grow up with a single mother, illustrating the compounding nature of inequality. Not only does lacking two parents' makes it harder for some kids to go to college and lead a comfortable life; in the aggregate, it also undermines social mobility and perpetuates inequality across generations. [71]

## Stage 2 of the Mimesis of Life

## Chapter 3

## Mimesis Summary

The Stage 2 Mimesis is the longest period of the Mimesis of Life. Stage 2 Mimesis begins when kids begin to associate with other kids and make their own decisions outside of the jurisdiction of their parents or caregivers. Decisions about the choice of friends begins the journey into Stage 2 Mimesis. The new discoveries kids encounter in early Stage 2 guides their progress throughout Stage 2 Mimesis. Stage two Mimesis is where the representation of the individual's reality begins is developed and stabilized. Stage 2 usually begins when the child starts nursery school or kindergarten and chooses their new friends. Experiencing new associations outside of their home and making decisions without parental supervision initiate the child into stage 2 mimesis construction. The child discovers individual differences in other kids. Not all kids are raised the same. Stage 2 Mimesis is where the individual accumulates life experiences and encounters a variety of phenomena. The individual's reality develops, molded, established and stabilized in Stage 2. As the child grows and attends elementary school their Stage 2 Mimesis continues. The accumulation of life experiences in Stages 2 Mimesis traditionally involves life experiences taking place in the following sequence of environments, life situations that require decision-making considerations.

        Nursery and Elementary School
        High School

Volume II of the Three-Fold Mimesis of Life

> Post High School, College. University. The workplace.
> Life Experiences and environments influence an individual's choices.
> What to do with your Life, embarking on a career, or attaining a job.
> Thoughts about your future. Job. Career. Responsibility.
> Marriage Family – More Responsibility, Becoming Parents.

Each of these life situations are sequenced as a continuum of Stage 2. While the environments a traditional sequence of an individual's experiences. Not all individuals experience the same sequence of life situations. Their life can have" hiccups" along the way or even take an individual off their desired course. Offsets are part of the individual's formation and construct of their reality. How an individual responds to unexpected situations, challenges, tragedy, disappointment, threats to their lived experience and other unexpected situations are part of their Mimesis construction. How individuals respond to life challenges is just as critical and important as how they respond to happy, rewarding and beneficial life experiences. Individuals will encounter undesirable life experiences. The good, bad and ugly life experiences feed the individuals Mimesis construction. How an individual responds to each circumstance they encounter is an emplotment/phenomena contributing to their narrative, and a reflection event in stage 3 Mimesis. Other challenges occurring in an individuals lived experiences may be: family, divorce, bills, debt, job stress, sickness, raising a child, parenting, periods of unemployment, disagreements, death of loved ones and friends, "automobile flat tires", are but a few. However, these challenges should not change the responsibility and duty required for constructing a positive Mimesis. Actually, challenges can act as motivators for positive mimesis construction. How an individual responds to life challenges determines their Mimesis.

The early stages of human development are critical to the formation of an individual's thinking and direction in Stage 2 Mimesis. How the individual is conditioned to respond to the experiences they encounter is seeded in Stage 1 Mimesis, then developed and stabilized in Stage 2 Mimesis. Stage 2 Mimesis is a further development of an individual's cognitive perspectives, thinking, and behavior. A youth who has developed a positive future-oriented cognitive bias starts out with an advantage. A positive future-oriented cognitive bias is a guide that gives direction to an individual's life. The early phase of Stage 2 is a continuation of the learning stage implanted in Stage 1. An individual's personality stabilization does not occur until middle adulthood. Personality traits and consistencies become noticeable during childhood. According to research published in the Journal of Psychology and Aging, lifelong personality stability is ambiguous, however, research evidence indicates that generally personality stability occurs from childhood (14 years old) to middle-late adulthood (30-40), The reality of an individual is noticeably stabilized from early middle adulthood to older age. During this period, other individuals notice that the individual's personality shows stability. Personality stability and the stability of an individual's reality constructs across the spectrum of their entire lived experiences.[72] How a person thinks is a determinant factor in how they behave. Mimesis stability in an individual's behavior is congruent with the representation of their reality. An individual's behavior is a display of their Mimesis. The Mimesis of Life is not a nebulous construct. There is reason and logic to an individual's Mimesis construct. The reality is that an individual can construct their own Mimesis with aforethought in their consciousness regarding how they respond to phenomena, and lived experiences they encounter, integrated with a future-oriented cognitive bias.

## CHAPTER 4

# Stage 3 of the Three-Fold Mimesis of Life: The Twilight Years of Life: Reflective

The one thing every human being on earth will experience is Stage 3 Mimesis; if they live long enough. Without doubt, everyone who is born will experience the Final Mimesis (death). The Final Mimesis of life is a guaranteed experience for everyone.

Stage 3 Mimesis is reflective. Individuals look back on their life and reflect, evaluate. Another aspect of Stage 3 Mimesis is that others reflect and look back on your life. How do you perceive yourself? How do others perceive you? Is there a common alignment between how you see yourself and how others see you? This is a significant aspect of the Three-Fold Mimesis. The input and involvement of others in the evaluation of your life. When your self-perception aligns with how others perceive you, then that is your reality. Mimesis is complete when the votes are in. Usually, the tally becomes final when the individual is deceased. An individual's Mimesis lasts as long as their memory in the thoughts and minds of their loved ones and friends. For example, the Mimesis of George Washington still exists. Same with Abraham Lincoln, Martin Luther King, Malcolm X, Albert Einstein, and Babe Ruth. People still refer to these individuals historically and in reference to their contributions. Their Mimesis is still in existence.

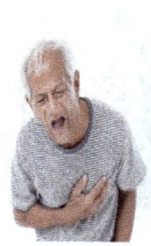  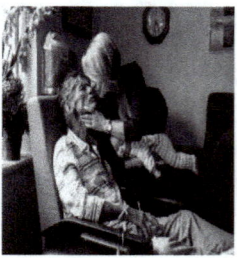

## Stage 3 - The Final Mimesis of Life

The Final Mimesis is, to some extent, independent of age; everyone dies. Normally, death occurs when people reach old age. The physical life expires. Consciousness and physical life on the earthly plane end. However, young people also die. Death can occur at any stage of Mimesis or at any time in life. Those who make better life choices are less prone to experience premature death, accidental death, or a senseless and needless ending to their life. It is at this stage, three in an individual's life Mimesis that others weigh in on your life. Your memory and legacy become a testament to how you have lived your life and how your Mimesis was constructed. Stage three Mimesis lasts as long as people have a memory of the individual. From the photos, it is evident why one reference to the state of death is "Laid to Rest".

The concept of death makes people experience uncertainty because, in many cases, they do not know when and how death will occur. Death is the only life event that is inevitable and unavoidable. No form of human hope, behavior, motivation, money, or desire will deter the final Mimesis. Death creates extraordinary anxiety, resulting in terror to human life. Death anxiety is a result of people living in the shadow of death.[73] There is evidence that a positive Mimesis construction allows individuals to mitigate the anxiety and depression resulting from the awareness of death and the experience of dying.[74] Religious and spiritual beliefs are often catalyst that mitigates the fear of death and allows people to deal with the final Mimesis in a natural way. The final Mimesis is an extension of life rather than the end of life.

A positive self-reflecting Mimesis can help people mitigate common realities; such as aging; thoughts of death, death anxiety, and death fear. When individuals feel good about the way they have lived their life, research indicates that it is a factor in mitigating death anxiety. According to research, having social curiosity redirects and diverts a senior person's focus from death anxiety and thoughts of life ending. *Social curiosity is an interest in obtaining new information and knowledge about the social world. Social curiosity increases one's ability to adapt and to survive. It deflects thoughts from life ending to active participation in discovering more about oneself, the world they live in and others.* Social curiosity enables individuals to appreciate themselves and others. Focus is on the positive aspects in an individual's life instead of the mundane. Social curiosity leads people to believe that after physically dying some valuable aspects of themselves will continue to exist, literally, such as in heaven, or in the memories of individuals they touched, or symbolically through gifts or imprints they made on society, or symbolically, such as self-prolongation through their children or their eternal achievements.[75]

Our contemporary perspective when confronted with death as an unavoidable reality can contribute to unsustainable social practices such as increased material acquisition and behaviors reminiscent of their younger years, sexual pursuits, drug use, alcohol use, and overextending the physicality of the body. The ways society offers to alleviate the anxiety of death are avoidance and distraction especially through consumption. Consumption (extreme experiences, materially, or gluttony), in this regard, theoretically, individuals have a tendency to raise individual self-esteem in order to strengthen the unrealistic attempt to avoid the inevitable. However, this behavior does not dismiss the reality that you will die. Disregarding our vulnerability can blind individuals to the necessity to maintain a life-supporting ecosystem and other society-sustaining phenomena. When death anxiety projects selfish behavior, it can have a negative impact on the individual, on others, and the society, in general. Ironically, exaggerated consumption in Stage 3 Mimesis can be counterproductive and quicken the final Mimesis. Constructing our Stage 3 Mimesis in terms of embracing human mortality with a positive perspective based on behaviors that are fulfilling and gratifying at that time in the life cycle can have mitigating results. One should not suppress the reality of death. The ability to deal with reality is a strength. Integrating one's present and future positive lived experiences with the awareness of reality gives a healthy balance to stage three Mimesis.

Organizing Stage three behaviors around the concept of lived experiences that develop and maintain the quality in relationships and focus on appreciating life. Value what we have instead of focusing on our fears. An individual can appreciate and feel gratitude if their Mimesis construction has been positive; the individual has much to appreciate. Consciously cultivating social relationships and reassessing values by focusing on meanings and projects that transcend individual selfishness. The point is that, when we are confronted with an unavoidable situation, an inevitable reality, how do we deal with it? With fear and anxiety that will cause additional distress over time, we have on earth or in a manner that will give us and those around us a meaningful and valuable perspective when reflecting. [76]

## CHAPTER 5

# Social Psychology and Performative Interventions in Human Systems

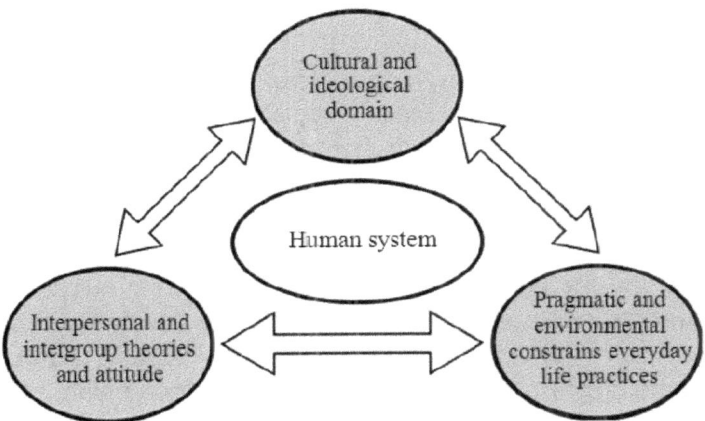

This chapter addresses the concept of systems that contribute to the development of Human System and that influence Mimesis development. On a micro-basis, the concept of behavior change is palatable. On a macro-basis, the concept of human behavior change is a dilemma. A rehabilitated criminal becoming a good citizen, is reasonable to perceive. An individual's Mimesis can take on a dynamic quality by changing environments, associations or habits. However, on a macro basis, eliminating crime from a society is dreaming. To understand and interpret the concept of development and changes in the human system and in the case of a society human systems, one has to accept the fact that a significant factor in human development is influenced by the environment in which people exist. Human systems are more susceptible to change when people change their environment or associations. Often that is challenging. Consider immigrants coming from another country to the United States. While they strive to fit into American culture, they are also resistant to changing their cultural traits (language, religion, etc.) and practices ingrained in their nature.

According to Nencini, Meneghini & Prati (2015) human beings act and react to social phenomena based on what it means to them or what meaning they attach the in their life. How their developmental process and Mimesis construct interprets the value of the phenomena to their life. Furthermore, Nencini, Meneghini & Prati (2015) advance that all human reality, at least in the universe of symbols and meaning, is generated by the interaction between individuals. This not only entails human interactions but the set of societal norms and values that are mutually known, believed, presupposed, or taken for granted by the individuals in the society with common environmental circumstances.[77] The Mimesis construction individuals develop is affected by human interactions and generally accepted societal norms and values. In most societies, the majority or dominant groups challenge sub-groups of individuals because the social constructionism is weighted in favor of those in power.

Groups that are in power attempt to control predict and assess the "real" reality of individuals in the society, even for those who are "out of power". In democratic countries; (America (Democrat and Republican parties), Canada (Liberal and Conservative parties), England (Labor and Conservative parties), France (Popular Front, Ensemble, National Rally and Republican parties), Germany (Christian Democratic Union, Christian

Social Union and the Social Democratic Parties)), there is more than one political party. Each party has different political ideologies that rule during their term in power. In democratic countries, the party in power can shift from election to election. In other countries, such as North Korea, China, which have a dictatorship or one-party system of government, the citizens are subject to follow the rules/laws of the dictatorship in power. Both systems of government, democracy and dictatorship, are heavily based on symbolic interactionism, and a social constructionism that posits human reality is neither objective nor universal but rather a social construction that exists in relationship with others. Within each of these respective environments, individuals construct their Mimesis. The differing environments affect the different way individuals in their respective environments relate to each other, organizations, and institutions in their society. Different norms and values are also prevalent in each environment. However, there are also Universal norms and values that all individuals are aware of and, for the most part, determine the order in a society. Laws deterring physical harm to others, stealing, property laws, laws preserving civil obedience, and laws that give order to ongoing society interactions. On an international basis, Mimesis construction varies due to environmental and societal differences.

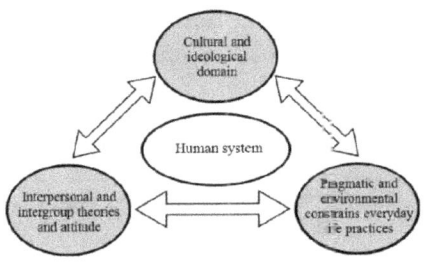

The graph depicts the construct of the Human System and the factors that influences and contributes to the Mimesis construction of human beings in their respective society. Generally, accepted cultural and ideological norms and values (Societal beliefs and generally accepted values) affect the human system. The impersonal and intergroup theories and attitudes contribute to human values (cultural beliefs). Additionally, the group in power under the domain of maintaining civil order establishes the environmental and pragmatic constructs. All of these systems can vary depending on the government party in power, the Government Domain running the country, and depending on how the power shifts occur. The human system is configured as a social construct influenced by an environmental, societal, and cultural context of beliefs, habits, norms, and values. The practical concept of human systems is manifested by the interaction of various roles in society. Every established role in the human system contributes to its creation and maintenance of society.

The ideology of the top manager or company president affects the behavior of the employees and the customers. The patient/doctor relationship is determined by established human systems in a health institution. The relationship between a government and the citizens in a society influences the way human systems and the Mimesis of citizens are manifested and constructed. Thus, the concept of human intervention/behavior change from an individual perspective is reasonable. From a society perspective, the practicality is not to remove problems, but to generate new, effective narratives for human interactions that mitigate problems.[78]

Volume II of the Three-Fold Mimesis of Life

CHAPTER 6

# Social Media Effect on the Child Developmental Process and on Society

Society evolves. One of the most significant evolutionary developments of the $20^{th}$ and $21^{st}$ centuries is digital technology and social media. Social media has influenced the construct of the Human Systems and the factors that influence and contribute to the Mimesis construction of human beings in their respective society. The impact of Social Media rivals, if not surpasses, the invention of the automobile, airplane, the printing press, or the telephone. Actually, technology has changed the entire characteristics of communication, from home landline phones/pay phones on the street to cell phones that give humans 24/7/365 ubiquitous communication access. Social media has eliminated the boundaries that previously separated individuals geographically and connects people worldwide. Digital technology plays a significant role in practically every area of life, including business, education, economics, religion, social interactions, transportation, and any other area of life one can think of. The one most significant impact on the lives of teenagers and young people is social media. Social media has a socializing impact on the lives of teenagers and young people. For this reason, understanding social media is important. Understanding its impact on kids to manage the influence and use of social media, void of the permeating dysfunctions, is important. Integrating social media into a child's socialization in ways that contribute to the child's developmental process can be a valuable learning tool, but it is the parent's responsibility to manage the child's usage of social media.

## Beginning of Social Media

Social media began in 1997 with an early platform called SixDegrees. Six years later around 2003 MySpace was born. No book written about these contemporary times about the human developmental process, especially with teens, would be complete without including a discussion about social media and its impact on Society and young people in particular. The social platform, SixDegrees was a pioneer in the social media arena. The SixDegrees platform allowed users to create profiles, add friends, and connect with others. MySpace, launched in 2003, quickly gained popularity and became one of the first social media platforms to reach a million monthly active users. MySpace had a significant influence on technology, pop culture, and music, and more so, unveiled the power of online social interaction. Other online social platforms like Facebook, YouTube, Twitter, Instagram, and TikTok emerged, further solidifying social media's place in daily life and shaping the way people connect online and with each other.[79]

Social media encompasses a wide range of online platforms where users can connect, share information, and build communities on many different topics. Connections based on text communication, sharing images, and videos. Platforms and user groups were created based on many areas of human interest: cooking, parenting, fashion, sports, and chess. User groups with common interests are online for almost any area of interest. These platforms have become an important and institutionalized part of modern society, used by billions of people globally daily. They offer opportunities for socialization, communication, and information sharing. On the other hand, there are online sites that have a potentially negative impact on individuals and society. Popular platforms include Facebook, Instagram, Twitter/X, TikTok, YouTube, LinkedIn, and WhatsApp.[80] These sites are a source of primarily wholesome media content and give the users a convenient source to communicate; however, even in these sites, negative influence can lurk. Child solicitations, drug dealers, fraud artists, and other culprits of dysfunction seek to negatively solicit and influence users.

Social Media's impact on society has been that of a double-edged sword. It has been an asset and produced considerable benefits on one hand, but on the other hand, it has caused deep cuts and been a source of tragedy that has torn families apart. Whereas social media has become a vast, informal network of online venues – public and private, paid and free subscription, large and small. Online venues are used for numerous purposeful tasks. Such as:

| | |
|---|---|
| Family Celebrations | Political campaigns |
| Charitable promotions | Health support groups |
| Sports Clubs | Prayer circles |
| Hobbies and occupations | Fundraisers |
| Information source on many topics | Personal Expression and Creativity |
| The Creation of Millions of Jobs | Facilitates Professional Networking |
| Creates Business Opportunities | |
| Facilitates Education Opportunities | Opens Marketing Channels |
| | Enhances Social Awareness |

In addition, social media is a way for people to keep in touch with each other, and more conveniently, stay in touch when distance separates them.[81] Social media has emphasized aspects of human behavior that reinforce the fact that people need to talk with like-minded people, people need "controversial" interaction with each other, people need to share life's joys, interests, and grief, people have a need to spending time with each other (albeit in the unconventional sense), and people want to watch each other. Technology advancements and the social media revolution have changed society and the way people interact with each other and with the world around them, personally, socially, and professionally.[82] Digital technology has become so widespread and invasive that "social media interaction, wearable devices, mobile applications, and pervasive use of sensors have created a personal information ecosystem for gathering traces of individual behavior" and, in some cases, detailed information on human activity. [83]

Advances in artificial intelligence (AI) can simulate human experiences that rival reality. The potential of technology to "immortalize ideas, reasoning, and behavior" raised intriguing, exciting, ambiguous, and frightening perceptions about the use of technology. The rapid growth of technology causes debates on the benefits of technology advancement, as well as debates on the ethics and morality of the directions technocrats are taking in technology. Technology seems to be innovating and moving faster than most people can keep up. Most people are uninformed about finding ways to integrate new technology positively and comfortably into their lifestyles. The real concerning issue is what impact will the expanding dominating applications of technology have on society, on people, and especially on the youth? Insight into this single issue will shed considerable light on the direction technology should take, how it should be used, and how it should be managed.

## The Emergence of Social Media

For all practical purposes, social media was born in 1997 with the launch of SixDegrees by Andrew Weinreich. He is considered the "father of social networking." SixDegrees' was a social platform for email connection links and basic networking. Individuals with shared interests were able to expand their people connections and sphere of influence without geographical limitations. However, in 1997, the technology evolution lagged considerably behind the ideology of social connections. Weinreich sold SixDegrees for $125 million in 1999. Social media technology, however, took off in the new millennium, and the 21$^{st}$ century saw not only an expansion in social media platforms but also a development in technology to facilitate the growing social media marketplace. Social media accelerated its growth to foster connections, build communities, and provide access to information.

With the invention of social media, ideologies to utilize this powerful phenomenon also grew. Applications sprang up in all areas of society: education, business, hobbies, advertising, influencing, and many more. The online platforms found applications in almost every area of society and human interest. Students can receive degrees via online application. Businesses operate by selling products online. Consumers shop online. Online education has opened the door, facilitating the learning of students who are working and do not have time for campus attendance. For individuals who have families, online education has made it convenient to get a college education. While online platforms have made shopping for goods and services convenient, it has also been a factor in the reduction of malls and the foot-traffic retail marketplace. Consumer shopping habits transformed from shopping at the mall and retail stores to the convenience of shopping without leaving the house. The grip social media has on our society is profound. Social media impacts how individuals in every generation connect, communicate, and consume information; influencing relationships, mental well-being, and society as a whole. Generation Z is the first generation that has grown up exclusively under the influence of technology. Generation Z (Gen Z) refers to the demographic cohort born between 1997 and 2012. As a result, Generation Z and the subsequent generation Alpha show the characteristic symptoms of a culture exclusively and significantly influenced by social media.

Individuals born from 2013 to the present are Generation Alpha. They are also the generation to be born exclusively in the 21st century. Their lived experiences are entirely technology influenced. Forty-three percent, (43%) of Generation Alpha individuals have a tablet by the age of six and 58% have a smartphone before they are 10 years old. The reason this information is important is that technology is integral to this generations Adolescent Brain Development, their early Childhood development, their health and social development, and more importantly their mental health. Generation Z and Alpha are proficient with the concepts of technology at an early age. This could give them the ability to process data at a quicker rate. It can also have a diminishing impact on the development of their social skills.[84] Astonishingly surprising is that "when it comes to product recommendations — almost half of Gen Alphas say that they trust their favorite influencers as much as their own family members". This is a blatant indication of how entrenched Generation Alpha is into social media. It is also an indication of a family value system in our society that is becoming dismantled.

## **Statistics on Gen Alpha Social Media Usage**

Gen Alphas are already showing trends in their social media usage. Research indicates the patterns of their social media usage are:[85]

- "More than 36 million children (ages 0–11) are active internet users, exceeding teen (ages 12–17) internet users by 11.6 million".
- "Sixty-five percent of Alphas ages 8–10 spend up to four hours a day on social media".
- "In 2024, 44% of Gen Alphas reported using Tiktok, outpacing their consumption of traditional television (39%)".
- "More than 30% of Gen Alphas watch YouTube and YouTube Shorts more than two hours daily".
- "Two-thirds of 6–8-year-olds search for Roblox videos on the platform at least weekly".

Out of concern for the future of our children and for our society, the question has to be addressed:

## Is GEN Z and Gen Alpha Addicted to Technology?

The only lived experience Generation Alpha has is in a digital society, and most of the lived experiences of Generation Z are digitally influenced. According to experts, this raises a concern about the impact of "the resulting negative effects on their mental, emotional, and behavioral development. There is even some concern about a potential dependency, or even addiction, to social media and technology for these kids".[86] Behavioral research indicates that exposure to "short, flashy video content across multiple screens, data suggest Gen Alphas struggle with dwindling attention spans, showing disinterest in activities that don't involve screens. Unfortunately, shortened attention spans can affect classroom behavior and learning retention".[87]

While conclusions drawn about long-term outcomes have a limited data analysis for drawing empirical conclusions, data show that "kids born during or around the COVID-19 pandemic are already showing delays in social and emotional development. A child's earliest years, from birth to age 5, are critical to establishing healthy social and emotional behaviors".[88] COVID lockdowns 81 prevented interpersonal socialization, resulting in increased time on screens, even for the youngest Alphas. Many of these children are exhibiting delays in their interpersonal and social well-being, which will pose future challenges to teachers and caregivers.[89]

Social media is so entrenched in Generation Alpha that it supersedes the need to connect with family and friends. Social media connects and aligns lifestyle preferences and gives access to new discoveries among

Generation Alpha. The astonishing reality about Generation Alpha and the role of social media in their life is that "49% of Generation Alpha kids say they trust influencers as much as their own family and friends when it comes to product recommendations".[90]

The Pre-adolescent stage for kids is between the ages of 9-12. The pre-teen years are the time when the child experiences significant social and emotional growth. It is also during this period in the development process when a child's personality, sense of independence, and confidence are seeded. During this period, the child explores experiences outside of their family.[91] The child is also impressionable and influenced at this stage of their developmental process. They are learning at a fast pace. Almost everything they encounter is a new learning experience. Their frequent use of social media influences their developmental process.

The preadolescent is at a crucial stage between childhood and adolescence. It is a period of vulnerability also. Their body is changing, their mind is developing, and their social circle is growing. The psychological impact of this period of life change for kids needs to be constructively guided and molded positively. During this period, social media serves as a "sociocultural mediator that shapes self-perceptions".[92] The influence of social media platforms, at this stage of a child's life, serves as a sociocultural mediator that shapes self-perception. Much of the communication is via activities such as photo sharing and self-representation.[93] The integration of physical and virtual realities, in addition to the constant online activity, is predominant among Generation Alpha youth. Technology entirely influences their lived experiences.

To understand the impact social media has on society, consider the following information:

### Social networking services with the most users, January 2024 [94]

| # | Network | Number of users (millions) | Country of origin |
|---|---|---|---|
| 1 | Facebook | 3,049 | United States |
| 2 | YouTube | 2,491 | United States |
| 3 | WhatsApp | 2,000 | United States |
| 4 | Instagram | 2,000 | United States |
| 5 | TikTok | 1,526 | China |
| 6 | WeChat | 1,336 | China |
| 7 | Facebook Messenger | 979 | United States |
| 8 | Telegram | 800 | Russia |
| 9 | Douyin | 752 | China |
| 10 | Snapchat | 750 | United States |
| 11 | Kuaishou | 685 | China |
| 12 | Twitter | 619 | United States |

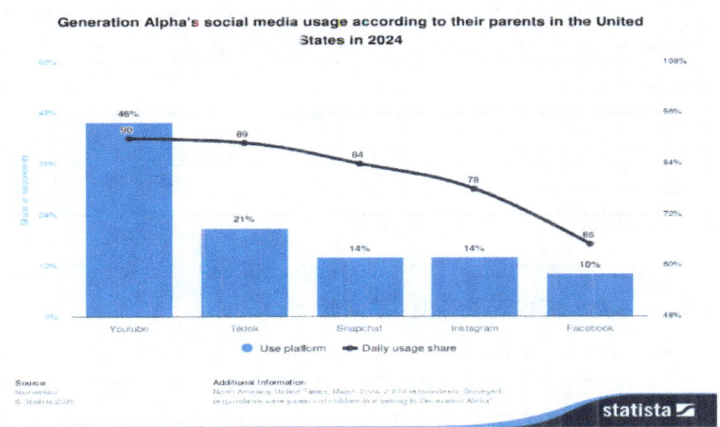

The chart shows the 2024 Generation Alpha's social media usage according to their parents in the United States.[95] The online social platforms they frequent (YouTube, TikTok, Snapchat, Instagram, and Facebook) influence, mold, and condition their body image, their self-esteem, their emotional intelligence, and their interpersonal relationships.[96] Generation Z (born between 1997-2012) and Generation Alpha (born between 2010-2024) are practically dependent on social media. Social media is their central platform for socializing and communicating with friends and for receiving information. Generation Z is referred to as the 'digital natives' - the first generation to be raised entirely within the internet and social media landscape.[97] The Millennial parents have a heavy reliance on technology and social media, so it is normal for them to encourage their kids to develop a technology interest. Millennial parents frequently share their family life on social media. There is concern that the frequency of social media usage by Generation Z and Generation Alpha is a factor in the diminished social skills of Generation Z and Generation Alpha.[98]

A research study on social media and Generation Alpha was published in the International Journal of Adolescence and Youth by Piccerillo, Tescione, Iannaccone, and Digennaro in January 2025. The article is entitled "Alpha generation's social media use: sociocultural influences and emotional intelligence". The findings of the research study were: [99]

- There is a high social media usage among preadolescents (Generation Z).
- There is a need to investigate and understand the social media effects on the mental health and sociocultural attitudes of this Generation
- There is a positive correlation between social media use and both addiction and internalization of beauty standards.
- There is a need for early interventions by caregivers to mediate a healthy frequency of social media usage among Generation Z youth.
- The addictive patterns of social media showed no difference between the impacts on genders.
- Excessive social media usage negatively impacts emotional intelligence.
- Furthermore, the research provides insights for educators, mental health professionals, and policy makers into the need for strategies and ideologies that mitigate the addictive impact of social media on these youth.
- There is a need to promote healthier lifestyle behaviors by moderating the use of online activity and social media usage among Generation Alpha.

Volume II of the Three-Fold Mimesis of Life

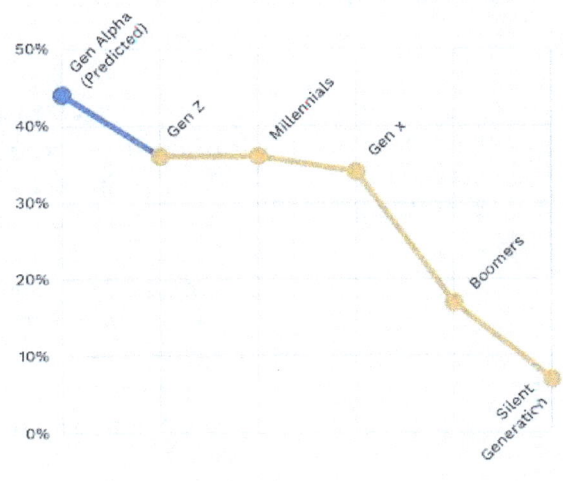

Online, internet, and social media usage among the generations.[100]

Gen-Alpha parents (Millennials) are technology savvy, open to new "digital trends", and often motivate the technology development of their children. The trend will grow. In the US alone, a new Gen Alpha baby is born every 9 minutes and statistically the Baby Boomers within four years. This and future generations are prone to be technology savvy and literate. Future projections indicate a severe change in cultural traditions, led and influenced by technological pursuits.

# Social Media Usage among Youth and Teens

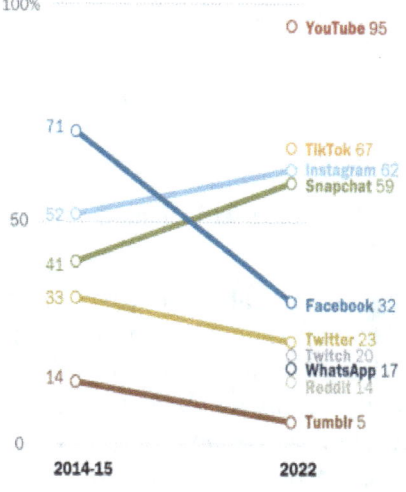

**Since 2014-15, TikTok has arisen; Facebook usage has dropped; Instagram, Snapchat have grown**

*% of U.S. teens who say they ever use any of the following apps or sites*

Note: Teens refer to those ages 13 to 17. Those who did not give an answer are not shown. The 2014-15 survey did not ask about YouTube, WhatsApp, Twitch and Reddit. TikTok debuted globally in 2018.
Source: Survey conducted April 14-May 4, 2022.
"Teens, Social Media and Technology 2022"

**PEW RESEARCH CENTER**

The Majority of teens use YouTube, TikTok, Instagram and Snapchat. YouTube is the platform most commonly used by teens, with 95% of those ages 13 to 17 saying they have ever used it, according to a Center survey conducted April 14-May 4, 2022, that asked about 10 online platforms. Two-thirds of teens report using TikTok, followed by roughly six-in-ten who say they use Instagram (62%) and Snapchat (59%). Much smaller shares of teens say they have ever used Twitter (23%), Twitch (20%), WhatsApp (17%), Reddit (14%) and Tumblr (5%). [101]

The chart below represents a demographic breakdown of social media usage among teenagers.[102]

**Teen girls are more likely than boys to use TikTok, Instagram and Snapchat; teen boys more likely to use Twitch, Reddit and YouTube; and Black teens are especially drawn to TikTok compared with other groups**

*% of U.S. teens who say they ever use each of the following apps or sites*

|  | YouTube | TikTok | Instagram | Snapchat | Facebook | Twitter | Twitch | WhatsApp | Reddit | Tumblr |
|---|---|---|---|---|---|---|---|---|---|---|
| Total | 95 | 67 | 62 | 59 | 32 | 23 | 20 | 17 | 14 | 5 |
| Boys | 97 | 60 | 55 | 54 | 31 | 24 | 26 | 17 | 20 | 4 |
| Girls | 92 | 73 | 69 | 54 | 34 | 22 | 13 | 18 | 8 | 6 |
| White | 94 | 62 | 58 | 59 | 32 | 20 | 20 | 10 | 16 | 5 |
| Black | 94 | 81 | 69 | 59 | 34 | 31 | 18 | 19 | 9 | 4 |
| Hispanic | 95 | 71 | 68 | 52 | 32 | 28 | 22 | 29 | 14 | 6 |
| Ages 13-14 | 94 | 61 | 45 | 51 | 23 | 15 | 17 | 16 | 8 | 3 |
| 15-17 | 95 | 71 | 73 | 65 | 39 | 29 | 22 | 18 | 19 | 7 |
| Urban | 95 | 71 | 70 | 58 | 40 | 28 | 15 | 29 | 13 | 6 |
| Suburban | 94 | 64 | 61 | 58 | 24 | 24 | 24 | 16 | 17 | 5 |
| Rural | 95 | 67 | 58 | 52 | 43 | 19 | 17 | 11 | 11 | 5 |
| *Household income* | | | | | | | | | | |
| < $30,000 | 93 | 72 | 64 | 30 | 44 | 26 | 17 | 19 | 10 | 4 |
| $30K-$74,999 | 94 | 68 | 62 | 37 | 39 | 24 | 19 | 19 | 13 | 7 |
| $75,000+ | 95 | 65 | 62 | 30 | 27 | 22 | 21 | 17 | 16 | 4 |

Note: Teens refer to those ages 13 to 17. Not all numerical differences between groups shown are statistically significant. Those who did not give an answer or gave other responses are not shown. White and Black teens include those who report being only one race and are not Hispanic. Hispanic teens are of any race.
Source: Survey conducted April 14-May 4, 2022.
"Teens, Social Media and Technology 2022"

**PEW RESEARCH CENTER**

The majority of these kids use social media on a daily basis, making social media habitual and a significant factor in their socialization process. More than 50% of teens indicate it would be difficult to give up social media. There is an irony in the teen psychology about social media.[103] Most teens indicate that social media makes them feel more connected to friends. Yet 20% of teen social media users believe that social media is detrimental to their mental health. [104]

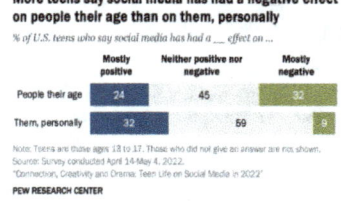

Thirty-two percent (32%) of teenagers say social media has a negative effect on people their age, while 9% say this about social media's effect on themselves.[105] This survey was taken in 2022.

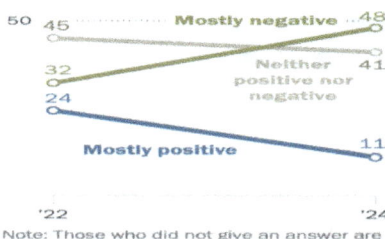

A survey taken in 2024, two years later, indicates that 48% of teens say social media is harmful to people in their age group. However, when it comes to a person's evaluation of the effect of social media on their own lives, 14% say social media has not affected them.

Parents are more concerned about the social media usage impact on the mental health of their teens than the teens are themselves.

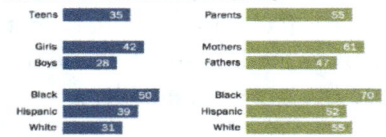

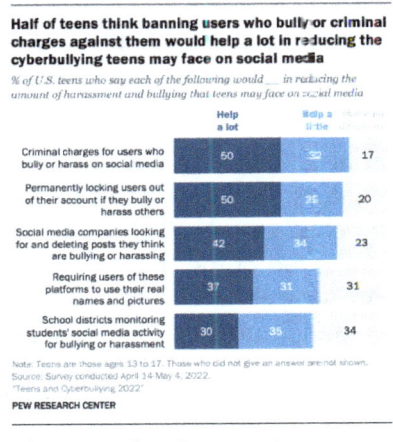

The facts is that social media is a controlling factor in the lives of teenagers requires that social media, like any other activity teens become involved, be monitored by parents and even government authorities. Social media companies have done a poor job of monitoring themselves. Their motive is to gain as many users as possible, not to dissuade users from their sites.

Teens, themselves, support banning users from social media who abuse the media. People who use social media in unsavory ways, should be banned according to teens. Users who exhibit behaviors such as bullying, banning, and harassment should be banned. Teens also believe that people should not hide behind social media and be required to show their real names and pictures. Three-in-ten teens say it would help a lot if school districts monitored students' social media activity for bullying or harassment. [106]

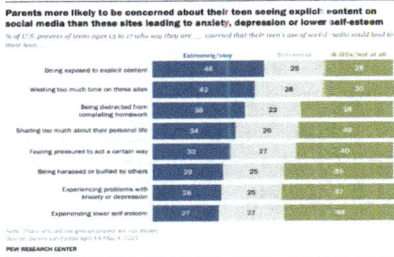

Parents have significant concerns about their kids using social media. The diagram to the left indicates parental concerns. [107]

Generation Alpha, Millennials and Generation Z use social media on a daily basis, however, Millennials, being older their social media usage is more matured, professionally and lifestyle focused.

One of the most problem users of social media among older adults are predators. Older men targeting minors and young teens for abusive and illicit purpose. Sextortion is a financially motivated crime that "involves an offender, who poses as someone else, coercing a minor to create and send sexually explicit images or video. An offender gets sexually explicit material from the child, and then threatens to release that compromising material unless the victim produces more. These offenders are seeking sexual gratification".[108] Sexual predators use social media online platforms to connect with young teens for sexual purpose. "According to the Federal Bureau of Investigations, 500,000 predators are scrolling through social media platforms, gaming apps and chat rooms, trying to connect with kids. Their targets are mostly age 15 and younger".[109]

## The Gloom and Doom of Social Media

The impact of social media depends on how it's used and the individual's ability to navigate its potential pitfalls; raises cause to argue that strong action needs to be taken to control and monitor social media because of the numerous and harmful negative impact it has had on society and especially our youth. The aforementioned downsides of social media do not come close to the extreme negative impact social media, has on society and on our children; including bullying, body shaming, harassment, abusive behaviors, murders and suicides.

## Negative Consequences of Social Media

- Concerns include the spread of misinformation, social media used as a breeding ground for cyberbullying, online harassment and potential for addiction.
- The spread of misinformation (fake news) and false information can have real-world consequences, affecting public trust and decision-making.
- Social media leads to serious emotional and psychological problems and consequences.
- Mental health issues develop when excessive social media use is linked to increased rates of depression, anxiety, loneliness, and feelings of inadequacy.
- Social media platforms are designed to be engaging and addictive, potentially leading to a decline in productivity and real-life interactions.
- Social media usage has led to numerous cases of identity theft and fraud. Swindling and cheating people out of money has been a major focus of internet criminals. The swindling schemes are numerous. Dating site deception, cryptocurrency fraud schemes, false imposter schemes, charity benefit schemes and more are just a few false presentations internet criminals use to steal money from unsuspecting individuals.
- Social media platforms collect vast amounts of user data, raising privacy concerns and potential risks of misuse.

There are aspects of social media use; the consequences thereof can be subtle. There is a much darker side of social media that has been responsible for murders and suicides. Mark Zuckerberg, President of Facebook, and other social media executives testified in from of congress in defense of social media. The outcome of his testimony presented the metrics of cost and Facebook practical issues pitted against the lives of Children.[110] There is no defense when it comes to the suicides and murders caused because of social media usage. It is true that social media is only a tool and the real culprits are people. The same argument applies to guns. Guns are only tools and the person using the gun is the culprit. However, if social media or the gun is a catalyst in the tragedies that occur in society, they, the catalyst, should be controlled. Politicians fail to enact consequential gun laws and people constantly are killed, especially kids. Kids killing kids should be grounds enough to enact serious gun control laws but the bribes politicians receive from gun lobbies are all they need to fail in their public duty.

A private school principal testified that everyone on his campus, students and teachers were much happier and more productive since the school banned phones.[111]

Depressive symptoms and suicidal thoughts for students grades 9 thru 12 have increased have increased since the oncoming of digital technology and social media. Social Media is not totally to blame for dysfunction among youth because there are many aspects of society that cause depression for young people. For example:[112]

**RACISM** – Contributes to depressive symptoms among targeted individuals

"Experiencing racism was 2–3 times more prevalent among students from all marginalized racial and ethnic groups compared to White students. Black students were more likely than White and Hispanic students to report unfair discipline at school.

High school students who experienced racism had a higher prevalence of poor mental health, suicide risk, and substance use".

"Especially impacted are: female students; lesbian, gay, bisexual, transgender, queer/questioning (LGBTQ+) students; and students from marginalized racial and ethnic groups".

"Having recognized the problem with teen depression, authorities have made minor headway in addressing the issues. The following information indicates progress from 2021 to 2023".

"The percentage of students overall who had persistent feelings of sadness or hopelessness (42% to 40%)".

"The percentage of female students who felt persistently sad or hopeless (57% to 53%) and who seriously considered attempting suicide (30% to 27%)".

"The percentage of Hispanic students who: felt persistently sad or hopeless (46% to 42%), experienced poor mental health (30% to 26%), seriously considered suicide (22% to 18%), and made a suicide plan (19% to 16%)".

"The percentage of Black students who attempted suicide (14% to 10%) and who were injured in a suicide attempt (4% to 2%)".

Again, without placing the total blame on social media for the rise in teenage problems and depression, Teenagers themselves reported the following:

"More than three-quarters of students reported frequent social media use. Frequent social media use was associated with":

- A higher prevalence of being bullied.
- Feelings of sadness and hopelessness.
- Suicide risk among students.

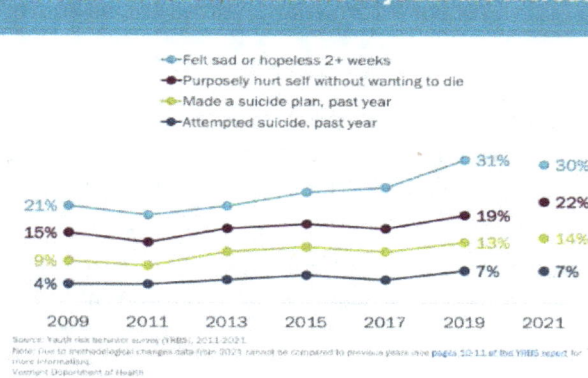

The "Youth Risk Behavior Survey" by the Vermont Department of health advances that social media could be a contributor to the increase in attempted youth suicide from 4% to 7%. This means an extra 3% per year could have been prevented, if the social media culture were constructed without negative phenomena. Vermont's teenage population is about 60,000, so 3% translates to 1,800 extra attempted suicides per year.[113]

There is significant reason to believe that the internet and social media has influence in suicide-related behavior among youth. Among adolescents and young adults, suicide is the second leading cause of death. Approximately 46,000 people died by suicide in the United States in 2020, according to the Centers for Disease Control and Prevention (CDC). "People die by suicide at a rate of 14 per 100,000". The suicide rate for children and young adults, 10 to 24, is 10.7 per 100,000.[114] "These groups have a high likelihood of encountering suicide-associated content on the Internet". According to the CDC, the suicide rate for male teens increased 31 percent between 2007 and 2015, and female teen suicides hit a 40-year high in 2015.[115]

In 2020, people in the US had a suicide rate of 13.5 per 100,000. Suicide accounted for 48,183 deaths in the United States in 2021. Suicide rates increased by 30 per cent from 2000-2018 and declined in 2019 and 2020. The facts and research indicate there is a causal relationship between social media usage and the increase in the suicide rate among teens. The purpose is to highlight the fact that social media usage is habitual among kids during the period they are most vulnerable. The integration of social media into the lives of teenagers and kids, factors into the suicide rates, according to research. Without drawing implications, these are the facts. [116]

The US population suffers 15,000 suicides per year, caused in aggregate by cyberbullying and harassment over smartphones and social media. Conservative reasoning estimates that approximately 2,000 extra teen suicides occur annually in the US due to social media and smartphone cyberbullying and harassment.[117] "There is increasing evidence that the

Internet and social media can influence suicide-related behavior".[118] In a search of 373 suicide sites, using the search terms such as (*suicide, suicide methods, how to kill yourself,* and *best suicide methods*), 11% of the sites returned in the search encouraged suicide.[119] For a kid or youth who is depressed and troubled, these sites contribute to suicide as a remedy for dysfunction when the person should be seeing a mental health professional. While most are suicide prevention sites, the fact that sites encourage suicide is concerning. "Both the U.S. Surgeon General and the American Psychological Association have issued advisories about the potential harms of social media for teenagers. Yale experts provide a guide for concerned parents".[120] The United States Office of the Surgeon General also published a "Social Media and Youth Mental Health Advisory" that all parents of teenagers should read.[121]

Teenagers and adolescents are susceptible to suicidal behavior because they learn about the suicide behavior of others. At this stage in their lives, they are impressionable. The suicidal sites online exposing young people to dysfunctional ideologies might increase "suicidal idealization".[122] Social media contributes to suicide risk behavior as a medium for and a catalyst in Cyberbullying and cyber harassment. Cyberbullying is when a child or adolescent is threatened, humiliated, harassed, or embarrassed by another child or "by means of cellular phones or Internet technologies such as e-mail, texting, social networking sites, or instant messaging. Cyber harassment and cyber stalking typically refer to these same actions when they involve adults". Survey data collected between 2004 and 2010 indicate that lifetime, cyberbullying victimization rates ranged from 20.8% to

40.6% and offending rates ranged from 11.5% to 20.1% of internet usage among preteens and teens.[123] Victims of cyberbullying were almost 2 times as likely to attempt suicide as those who were not. These results also indicated that cyberbullying offenders were 1.5 times as likely to report having attempted suicide as children who were not offenders or victims of cyberbullying.[124] While cyberbullying may not be the sole predictor of suicide in adolescents or teens, it can increase the risk of suicide for vulnerable individuals. Feelings of vulnerability such as; isolation, instability, and hopelessness for those with preexisting emotional, psychological, or environmental stressors makes the propensity of their vulnerability more susceptible to suicide risk. The vulnerabilities among the youth in American society are not diminishing. They are increasing. The CDC portrays U.S. high school students as existing in a state of distress. An increasing number of students reported persistent feelings of sadness or hopelessness in 2021, including 57% of girls (up from 36% in 2011), 29% of boys, and 69% of LGBTQ+ students. [125]

Research considers Social Media usage to be a catalyst for many negative acts, behaviors, thoughts, and feelings among youth. This is especially true regarding the effect of social media on mental health and the well-being of vulnerable individuals. A considerable number of teens consider social media as a negative medium. Consider the following Statistics.[126]

Mental Health:[127]
- **Anxiety and Stress:** "56% of social media users feel anxious when comparing themselves to their friends, and 36% of UK adults believe social media worsens their stress levels".

- **Loneliness:** "64% of people say social media increases their feelings of loneliness".
- **Depression:** "Studies show that teenage and young adult users who spend the most time on social media have a substantially higher rate of reported depression".
- **FOMO (Fear of Missing Out):** "69% of millennials experience FOMO regularly".
- **Low Self-Esteem:** "Social media comparison can lead to a decrease in self-esteem and distorted body image".

Effects on Teens:[128]

- **Negative Impact:** "25% of teens view social media as having a negative effect".
- **Pressured to Post:** "Almost 37% of teens feel pressured to post content that will be popular and get likes".
- **Increased Unhappiness:** "Eighth-graders who spend over 10 hours on social media per week are 56% more likely to report being unhappy".
- **Girls' Experiences:** "Teen girls are more likely to say social media negatively impacts their mental health, confidence, and sleep".
- **Excessive Use:** "45% of teens say they spend too much time on social media".

Social Media Addiction:[129]

- **Addiction:** "53% of young adults feel like they can't control how much they use social media".
- **Withdrawal Symptoms:** "11% of adolescents show signs of problematic social media behavior, struggling to control their use and experiencing negative consequences".

Other Negative Effects: [130]
- **Cyberbullying:** "88% of teens have seen someone be mean or cruel to another person on a social networking site".
- **Harm to Relationships:** "48% of young adults say social media harms their relationships".
- **Spread of Misinformation:** "64% of Americans say social media have a mostly negative effect on the country".

Historically, Black youths have experienced lower rates of suicide and suicide attempts than their White counterparts have. However, recent data, surprisingly suggesting that suicide rates are increasing among Black youth as well. *"Current evidence-based reviews suggest that dialectical behavior therapy is the only well-established treatment against self-harm and suicide among youths. However, it is unknown whether current established treatments work for Black youths, because Black youths are rarely included in randomized controlled trials.* [131]

The stakeholders for health social media usage are youth, parents, caregivers, educators, policymakers, practitioners, and members of the tech industry. A symbiotic reality is that social media behaviors are symptomatic of a kid's lived experiences online and offline. The impact of social media on a kid's lived experiences is reflective of the youth's personal and psychological characteristics and social circumstances, which often intersect with the social media content the youth is drawn to. Teenage and adolescent social media behaviors often result from the pre-existing disposition of the individual's psychology, vulnerabilities, and the context of their socialization process.[132]

"Adolescent development is gradual and continuous, beginning from birth. Biological and neurological changes occur before puberty are observable (i.e., approximately beginning at 10 years of age), and lasting at least until dramatic changes in youths' social environment (e.g., peer, family, and school context) and neurological changes have completed (i.e., until approximately 25 years of age). Age-appropriate use of social media should be based on each adolescent's level of maturity (e.g., self-regulation skills, intellectual development, and comprehension of risks) and home environment". Adolescents mature and develop at different rates. It is uncertain when and at what age kids become vulnerable to the potential risks posed by social media".[133] Most likely the potential "risks are likely to be greater in early adolescence" because this is a period of greater biological, social, and psychological transitions, than in late adolescence and early adulthood. The child is in the discovery stage and vulnerable to risks that may not be in their best interest. Their social protective and defense mechanisms are not developed. It is during this period of vulnerability when the youth needs parental jurisdiction, monitoring, and guidance in directing the lived experiences of the child.

Research indicates that the internet platforms often reflect the perspectives of the individuals who build the technology. Further indications are that there is a considerable degree of racism built into social media platforms. "For example, algorithms (i.e., a set of mathematical instructions that direct users' everyday experiences down to the posts that they see) can often have centuries of racist policy and discrimination encoded. Social media can become an incubator, providing community and training that fuel racist hate.

The resulting potential impact is far-reaching, including physical violence offline, as well as threats to well-being.[134]

As stated and reiterated throughout this book, social media has become functional and significant in the developmental process of adolescents, pre-teens and teenagers. Often youth with symptoms of mental illness, such as adolescents with social anxiety, depression, or loneliness, may heavily on social media as an outlet for their conditions. The irony is that these individuals are often the victims of social media dysfunction. The role that social media plays in the socialization process makes it imperative for parents to become involved in their child's social media behavior from the outset of their social media usage. The need for Parental control, monitoring, and supervision is critical. A consistent review of a child's social interactions is necessary. Parents should monitor social media with their kids whether mentally healthy or mentally challenged. Parents should participate in their child's social media sessions. Engaging participative social media usage from the beginning of a child's social media use will condition and make the child comfortable with adult participation in their online sessions. The parent will also be better able to guide the child's social media behavior and monitor the sites they visit. The American Psychological Association recommends the following health advisories on social media use among adolescents.[135]

1. **"Youth using social media should be encouraged to use functions that create opportunities for social support, online companionship, and emotional intimacy that can promote healthy socialization"**

2. "Social media use, functionality, and permissions/consenting should be tailored to youths' developmental capabilities; designs created for adults may not be appropriate for children".

3. "In early adolescence (i.e., typically 10–14 years), adult monitoring (i.e., ongoing review, discussion, and coaching around social media content) is advised for most youths' social media use; autonomy may increase gradually as kids age and if they gain digital literacy skills. However, monitoring should be balanced with youths' appropriate needs for privacy".

4. To reduce the risks of psychological harm, adolescents' exposure should be restricted against sites and social media content that depicts illegal or psychologically maladaptive behavior. This includes content that instructs or encourages youth to engage in health-risk behaviors, such as self-harm (e.g., cutting, suicide), harm to others, or those that encourage eating-disordered behavior (e.g., restrictive eating, purging, excessive exercise) should be minimized, reported, and removed;[23] Moreover, technology that drives users to dysfunction and harmful content should be diverted and eliminated.

5. "To minimize psychological harm, adolescents' exposure to "cyber hate" including online discrimination, prejudice, hate, or

cyberbullying especially directed toward a marginalized group (e.g., racial, ethnic, gender, sexual, religious, ability status),[22] or toward an individual because of their identity or ally ship with a marginalized group should be minimized".

6. "Adolescents should be routinely screened for signs of "problematic social media use" that can impair their ability to engage in daily roles and routines, and may present risk for more serious psychological harms over time".

7. "The use of social media should be limited so as to not interfere with adolescents' sleep and physical activity".

8. "Adolescents should limit use of social media for social comparison, particularly around beauty- or appearance-related content".

9. "Adolescents' social media use should be preceded by training in social media literacy to ensure that users have developed psychologically-informed competencies and skills that will maximize the chances for balanced, safe, and meaningful social media use".

10. "Substantial resources should be provided for continued scientific examination of the positive and negative effects of social media on adolescent development".

Kids are Obsessed with Social Media. The online technology internet culture has become integral to the lives of pre-teens, adolescents and teenagers. The "cat is out the bag" the only way to deal with it now is for parents to control their child's internet behavior from the first introduction of their awareness and usage. Among their peer-group values, the worse thing a parent can do is take the child's cell phone away from them as punishment. [136]

It is not exactly quantified how much of the mental health issues experienced by teens is attributed to social media. However, since the inception and proliferation of social media dysfunction among teenagers mitigating actions have to be taken for the safety of the children. Educators indicate that, "Since social media took off as a popular phenomenon in the early 2000s, the rate of adolescent depression has significantly spiked. Between 2005 and 2017, depression among young people reportedly went up 52%. [137]

Volume II of the Three-Fold Mimesis of Life

# Conclusion

Everything is on the internet, and I mean everything. There is practically nothing, no subject, no content, no discussion, or no topic you can think of that is not online. The internet is a treasure trove of information. Anything you want to know, "Google it".

The term "Google" is the trademark name of a web search engine. The ubiquitous use of Google as a search engine has coined the term as an action synonymous with "search". Foxfire, Bing, Baidu, and Yahoo are but a few of over 25 different search engines, some of which offer distinct advantages over Google,[138] but Google being the most popular, induces the internet user community to habitually use the term "google" when searching for internet content. Because of market dominance, Google is the main search. Google's algorithms are general and require adult intervention and oversight to manage the content available to children. The google navigation algorithms do have built-in tools and features, such as Google SafeSearch and the Family Link application, to help manage search topics for children. SafeSearch filters explicit content, and it is automatically set to a stricter filtering level for child accounts managed by Family Link. Parents can use Family Link to control their children's search experience, while school administrators can enforce SafeSearch on school-managed accounts. However, these safeguards require adult intervention and management.

The internet produces and receives messages, disseminates messages, and evaluates and regenerates content. The internet is a Pandora's Box. It is out there is no getting it back in. The content put out there becomes the property of cyberspace. The internet has become fully functional and integrated into the developmental process of our society's youth. I would guess that child development psychologists, Freud, Piaget, Bandura, Bowlby, Ainsworth, Erickson, or Vygotsky never imagined a social force such as the internet coming into existence, not to mention the impact of the internet on the developmental process of children. Nevertheless, the basic principles of their developmental theories are still relevant. Concepts on the development and socialization process of children are still basic to the learning experience, and parents are critical to the molding process of children. However, factoring in technology creates potential for greater exposure to more information at an earlier age.

In general, the advancement of technology to accommodate online access, social media, and application usage has created societal benefits that have quickly become institutionalized and integral to American lifestyle behaviors. Online behaviors have made life easier and more convenient in the following ways.

- **Connection and Social Interaction:**
  Social media allows users to connect with friends, family, and other like-minded individuals, fostering a sense of community and belonging.

- **Use in Marketing:**
  Businesses utilize social media for marketing, brand building, and customer engagement.

- **Information Sharing and Access:**
  It provides a platform for sharing information, news, and opinions, making it easier to stay informed and engaged with current events.

- **Networking and Career Opportunities:**
  Social media can be a valuable tool for professional networking, job searching, and finding new opportunities.

- **Creative Expression and Entertainment:**
  It offers platforms for sharing creative content, engaging in online games, and experiencing diverse forms of entertainment.

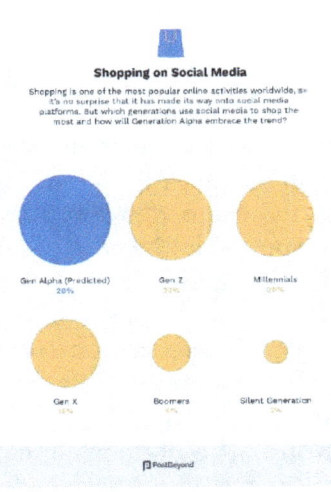

Online Shopping has become one of the fastest growing uses of online activity. Online shopping has grown so rapidly and widespread that the foot traffic in the malls and brick-and-mortar stores are facing increased competition from online retailers, leading to store closures and shifts in business models. The growth of e-commerce. Companies like Amazon, eBay and ASOS have changed the shopping behavior of consumers by making it possible to purchase goods from the comfort of their homes. [139]

One can make the argument that our society and our world has self-destructive tendency disguised under the framework of "Progress". For example, Nuclear energy has significant potential to contribute to the world societies. That is until it becomes a stock piled as nuclear arsenals, weapons with the potential for world destruction. Same for Social Media. What is it about human nature that they can devise detrimental and destructive use from inventions that have the potential for positive society contribution?

Here is where people miss the boat. The concerns of the Social Media Companies, investors and operators should be secondary to what is in the best interest of the society and especially the young people in the society. That the legalities allow social media to reign priority over what is in the best interest of our young people is a tragic flaw in America and the institution of social media. America is a country of free speech. However, while people have the right to express their opinions, there are numerous incidents where people are sued, and penalized for slander and making false statements. As one young woman put it, "We live in a society where viewpoints are more important than human decency.[140] In the case of the internet, retribution for cyberbullying and behaviors that motivate vulnerable individuals to harm themselves are not actively scrutinized or monitored. There are numerous articles, news stories and incidents of the role social media played about the deaths of preteens, teens and youth in America and throughout the world. The preteens, teens and youth are not the only victims of social media negativity and fatality. The parents and loved ones are victims who have to live with the suffering and the aftermath of social media tragedy encountered by their children.

Is it good to avoid social media? The answer to this question is "YES", if it makes you more negative, is a source of dysfunction in your life or induces a vulnerable user to commit self-harm. Social media is an outlet for many individuals to relieve their stresses, frustrations and heartbreaks, or a place to share their thoughts and feelings on the tragedies of the world. While it may make culprits feel better to unleash negatiity, it may not relieve those who are reading it. Sleep problems, attention problems, and feelings of exclusion among teenagers, is linked to excessive social media use. In a review of 36 studies concluded, there is a causal relationship between cyberbullying on social media and depression among children of all ages. Research indicates that more than 20% of teens have considered suicide. [141]

The real key to healthy online usage is the responsibility of parents to monitor their child's computer and social media usage. Parents need to monitor and control the sites their kids visit, the online friends they make and the interactions they have with their online connections. Without parental monitoring of the child's social media usage the risk are greater that the kids will encounter and become victim to the detriment effects of online predators and social media negativity. With parental involvement in the child's social media activity, the following rules are important to know.

1) **Moderation and Mindfulness:**
   To mitigate the negative effects of social media, it's important to use it in moderation and with mindfulness. This includes:

2) **Setting Time Limits:**
   Limiting time spent on social media platforms can help prevent addiction and promote a better balance in life.

3) **Being Mindful of Content Consumption:**
   Being aware of the potential impact of online content on mental health and emotional well-being can help avoid negative comparisons and unhealthy trends.

4) **Promoting Healthy Online Habits:**
   Developing positive online habits, such as engaging in constructive interactions, seeking out authentic content, and prioritizing real-life relationships, can help maximize the benefits of social media.

5) **Seeking Support When Needed:**
   If social media use is negatively impacting mental health or relationships, it's important to seek professional support and resources.

Social media is a powerful tool with the potential to both benefit and harm users. By using it responsibly and mindfully, individuals can harness its positive aspects while minimizing the potential negative consequences. There is no doubt that Social-Media and the digital age is here to stay. The direction, growth and impact of social media and digital phenomena are still in the development states. The maturity of digital interactions will be stabilized and improved by Generation Alpha, who themselves are not far from adulthood. Publishers, advertisers, and social platforms all will be focusing on how they will appeal to this generation, who, within four years, will be the largest generation in the world.

Research indicates that four out of five Gen Alpha kids significantly influence family purchases, important information for digital marketers. Smartphones will be commonplace with children at a younger age. Similar to getting a new toy. This early exposure to technology will shape and mold how Gen Alpha and future generations interact with the world. This prognosis makes it imperative that technology and social media take a more responsible role to ensure the content of their platforms have a positive influence on users and contribute positively to the lived experiences of platform users. Monitoring, mitigating, eliminating or at least considerably minimizing the content and effects of fake news, fake profiles, and fake reviews; social sites can guard against the negative impacts arising from decreased trust and the dysfunctional behaviors of cyberbullying, body shaming, harassment and trolling.

Volume II of the Three-Fold Mimesis of Life

# End Notes

1

Stages of Development. Introductory Psychology. Retrieved from: https://courses.lumenlearning.com/suny-hccc-ss-151-1/chapter/stages-of-development/

2

Stages of Development, Chapter 9. Lifespan Development. Psychology. Authored by: OpenStax College. Located at: http://cnx.org/contents/4abf04bf-93a0-45c3-9cbc-2cefd46e68cc@4.100:1/Psychology. https://courses.lumenlearning.com/suny-hccc-ss-151-1/chapter/stages-of-development/

3

Stages of Development, Chapter 9. Lifespan Development. Psychology. Authored by: OpenStax College. Located at: http://cnx.org/contents/4abf04bf-93a0-45c3-9cbc-2cefd46e68cc@4.100:1/Psychology. https://courses.lumenlearning.com/suny-hccc-ss-151-1/chapter/stages-of-development/

4

Stages of Development, Chapter 9. Lifespan Development. Psychology. Authored by: OpenStax College. Located at: http://cnx.org/contents/4abf04bf-93a0-45c3-9cbc-2cefd46e68cc@4.100:1/Psychology. https://courses.lumenlearning.com/suny-hccc-ss-151-1/chapter/stages-of-development/

5

Fitzgerald, E., Hor, K., & Drake, A. J. (2020). Maternal influences on fetal brain development: The role of nutrition, infection and stress, and the potential for intergenerational consequences. *Early human development*, *150*, 105190.

https://doi.org/10.1016/j.earlhumdev.2020.105190. Retrieved from:
https://www.ncbi.nlm.nih.gov/pmc/articles/PMC7481314/

6

Hertenstein, Matthew J. Social Referencing. Retrieved from:
https://www.depauw.edu/learn/lab/publications/documents/infant%20development/2010_Infant_development_Social_referencing.pdf

Hertenstein, M.J. (2011). Social Referencing. In: Goldstein, S., Naglieri, J.A. (eds) Encyclopedia of Child Behavior and Development. Springer, Boston, MA.
https://doi.org/10.1007/978-0-387-79061-9_2704.
https://link.springer.com/referenceworkentry/10.1007/978-0-387-79061-9_2704

Walden, T. A., & Ogan, T. A. (1988). The development of social referencing. *Child development*, *59*(5), 1230–1240.
https://doi.org/10.1111/j.1467-8624.1988.tb01492.x.
Retrieved from: https://pubmed.ncbi.nlm.nih.gov/3168639/

7

Nelson, C. A., Zeanah, C. H. and Fox, N. A. (2019). How Early Experience Shapes Human Development: The Case of Psychosocial Deprivation. Neural Plast. 2019; 2019: 1676285. Published online 2019 Jan 15. doi: 10.1155/2019/1676285. Retrieved from:
https://www.ncbi.nlm.nih.gov/pmc/articles/PMC6350537/

8

Kirsten Weir (2014). The lasting impact of neglect. American Psychological Association. Retrieved from:
https://www.apa.org/monitor/2014/06/neglect

9

Kirsten Weir (2014). The lasting impact of neglect. American Psychological Association. Retrieved from:
https://www.apa.org/monitor/2014/06/neglect

10

Ages and Stages of Development. California Department of Education. Retrieved from:
https://www.cde.ca.gov/SP/CD/re/caqdevelopment.asp

11

Borba, Michele (2022). Child psychologist: The No. 1 skill that sets mentally strong kids apart from 'those who give up'—and how parents can teach it. Retrieved from:
https://www.cnbc.com/2022/07/04/psychologist-shares-the-top-skill-that-sets-mentally-strong-kids-from-those-who-give-up-easily.html

12

JULIANA MENASCE HOROWITZANDNIKKI GRAF (2019). Most U.S. Teens See Anxiety and Depression as a Major Problem among Their Peers. *Pew Research Center.*
https://www.pewresearch.org/social-trends/2019/02/20/most-u-s-teens-see-anxiety-and-depression-as-a-major-problem-among-their-peers/

13

JULIANA MENASCE HOROWITZANDNIKKI GRAF (2019). Most U.S. Teens See Anxiety and Depression as a Major Problem among Their Peers *Pew Research Center.*
https://www.pewresearch.org/social-trends/2019/02/20/most-u-s-teens-see-anxiety-and-depression-as-a-major-problem-among-their-peers/

14

CDC. Anxiety and depression in children: Get the facts. Centers for Disease Control and Prevention. Retrieved from:

https://www.cdc.gov/childrensmentalhealth/features/anxiety-depression-children.html ; https://www.cdc.gov/children-mental-health/about/index.html?CDC_AAref_Val=https://www.cdc.gov/childrensmentalhealth/features/anxiety-depression-children.html

Menasce, Julianna (2019). Most U.S. Teens See Anxiety and Depression as a Major Problem Among Their Peers. *Pew Research Center*. Retrieved from:
https://www.pewresearch.org/social-trends/2019/02/20/most-u-s-teens-see-anxiety-and-depression-as-a-major-problem-among-their-peers/

Garber, J., & Weersing, V. R. (2010). Comorbidity of Anxiety and Depression in Youth: Implications for Treatment and Prevention. *Clinical psychology : a publication of the Division of Clinical Psychology of the American Psychological Association*, *17*(4), 293–306. https://doi.org/10.1111/j.1468-2850.2010.01221.x . Retrieved from:
https://www.ncbi.nlm.nih.gov/pmc/articles/PMC3074295/

15

CDC. Anxiety and depression in children: Get the facts. Centers for Disease Control and Prevention. Retrieved from:
https://www.cdc.gov/childrensmentalhealth/features/anxiety-depression-children.html

Menasce, Julianna (2019). Most U.S. Teens See Anxiety and Depression as a Major Problem Among Their Peers. *Pew Research Center*. Retrieved from:
https://www.pewresearch.org/social-trends/2019/02/20/most-u-s-teens-see-anxiety-and-depression-as-a-major-problem-among-their-peers/

Garber, J., & Weersing, V. R. (2010). Comorbidity of Anxiety and Depression in Youth: Implications for Treatment and

Prevention. *Clinical psychology : a publication of the Division of Clinical Psychology of the American Psychological Association*, *17*(4), 293–306. https://doi.org/10.1111/j.1468-2850.2010.01221.x . Retrieved from: https://www.ncbi.nlm.nih.gov/pmc/articles/PMC3074295/

16

ALEJANDRA CORTAZAR and FRANCISCA HERREROS (2010). Early Attachment Relationships and the Early Childhood Curriculum. *Contemporary Issues in Early Childhood Volume 11 Number 2 2010 www.wwwords.co.uk/CIEC*. Retrieved from:
https://journals.sagepub.com/doi/pdf/10.2304/ciec.2010.11.2.192

17

ALEJANDRA CORTAZAR and FRANCISCA HERREROS (2010). Early Attachment Relationships and the Early Childhood Curriculum. *Contemporary Issues in Early Childhood Volume 11 Number 2 2010 www.wwwords.co.uk/CIEC*. Retrieved from:
https://journals.sagepub.com/doi/pdf/10.2304/ciec.2010.11.2.192

18

EL Education. Characteristics of Primary Learner. Retrieved from https://eleducation.org/uploads/downloads/ELED-CharacteristicsofPrimaryLearners-0815.pdf
EL Education, 247 West 35th Street, 8th Floor, New York, NY 10001

19

Ages and Stages of Development. California Department of Education. Retrieved from:
https://www.cde.ca.gov/SP/CD/re/caqdevelopment.asp

20

Abdulaziz Al Odhayani, MD, William J. Watson, MD CCFP FCFP and Lindsay Watson, MA RMFT (2013). Behavioural consequences of child abuse. *Can Fam Physician. 2013 Aug; 59(8): 831-836.* Retrieved from:
https://www.ncbi.nlm.nih.gov/pmc/articles/PMC3743691/

WHO (2024). Child maltreatment. *World Health Organization.* https://www.who.int/news-room/fact-sheets/detail/child-maltreatment#:~:text=Overview,Estimates%20depend%20on:

21

Abdulaziz Al Odhayani, MD, William J. Watson, MD CCFP FCFP and Lindsay Watson, MA RMFT (2013). Behavioural consequences of child abuse. *Can Fam Physician. 2013 Aug; 59(8): 831-836.* Retrieved from:
https://www.ncbi.nlm.nih.gov/pmc/articles/PMC3743691/

Rizvi MB, Conners GP, Rabiner J. (2025). New York State Child Abuse, Maltreatment, and Neglect. [Updated 2025 Feb 19]. In: StatPearls [Internet]. Treasure Island (FL): StatPearls Publishing; 2025 Jan-. Available from:
https://www.ncbi.nlm.nih.gov/books/NBK565843/ ; https://www.ncbi.nlm.nih.gov/books/NBK565843/#:~:text=The%20World%20Health%20Organization%20(WHO)%20defines%20child%20abuse%20and%20child,Sexual%20abuse%20%5B1%5D

22

WHO (2022). Violence INFO: Child Maltreatment. *World Health Organization.* https://apps.who.int/violence-info/child-maltreatment/

23

ALEJANDRA CORTAZAR and FRANCISCA HERREROS (2010). Early Attachment Relationships and the Early

Childhood Curriculum. *Contemporary Issues in Early Childhood Volume 11 Number 2 2010* www.wwwords.co.uk/CIEC. Retrieved from:
https://journals.sagepub.com/doi/pdf/10.2304/ciec.2010.11.2.192

24

ALEJANDRA CORTAZAR and FRANCISCA HERREROS (2010). Early Attachment Relationships and the Early Childhood Curriculum. *Contemporary Issues in Early Childhood Volume 11 Number 2 2010* www.wwwords.co.uk/CIEC. Retrieved from:
https://journals.sagepub.com/doi/pdf/10.2304/ciec.2010.11.2.192

25

Sroufe, L.A. (2005) Attachment and Development: a prospective, longitudinal study from birth to adulthood, Attachment & Human Development, 7(4), 349-367.
http://dx.doi.org/10.1080/14616730500365928

26

Jane Anderson (2014). The impact of family structure on the health of children: Effects of divorce. Linacre Q. November, 2014; 81(4): 378–387.
doi: 10.1179/0024363914Z.00000000087. Retrieved from: https://www.ncbi.nlm.nih.gov/pmc/articles/PMC4240051/

27

Laura Broadwell (2005). Age-by-Age Guide on the Effects of Divorce on Children. Parents.Com. Retrieved from:
https://www.parents.com/parenting/divorce/coping/age-by-age-guide-to-what-children-understand-about-divorce/

### 28

DIVORCE STATISTICS: OVER 115 STUDIES, FACTS AND RATES FOR 2022. Retrieved from: https://www.wf-lawyers.com/divorce-statistics-and-facts/
Maria Indelicato (2021). In What Year of Marriage is Divorce Most Common. *Marriage. Com.* Retrieved from: https://www.marriage.com/advice/divorce/what-year-of-marriage-is-divorce-most-common/

### 29

Laura Broadwell (2005). Age-by-Age Guide on the Effects of Divorce on Children. *Parents.Com.* Retrieved from: https://www.parents.com/parenting/divorce/coping/age-by-age-guide-to-what-children-understand-about-divorce/

### 30

Wallerstein J.S., and Blakeslee S.. 2004. *Second chances: Men, women, and children a decade after divorce.* Boston, MA: Houghton Mifflin.

### 31

Caruso, E. M., Latham, A. J., & Miller, K. (2024). Is future bias just a manifestation of the temporal value asymmetry? Philosophical Psychology, 1–40. https://doi.org/10.1080/09515089.2024.2360673 ; https://www.tandfonline.com/doi/full/10.1080/09515089.2024.2360673#abstract

McCormack, T., Burns, P., O'Connor, P., Jaroslawska, A., & Caruso, E. M. (2019). Do children and adolescents have a future-oriented bias? A developmental study of spontaneous and cued past and future thinking. *Psychological research*, *83*(4), 774–787. https://doi.org/10.1007/s00426-018-1077-5 ; https://pubmed.ncbi.nlm.nih.gov/30159672/#:~:text=Abstract,bias%20may%20be%20task%2Dspecific.

**32**

Caruso, E. M., Latham, A. J., & Miller, K. (2024). Is future bias just a manifestation of the temporal value asymmetry? Philosophical Psychology, 1–40. https://doi.org/10.1080/09515089.2024.2360673 ; https://www.tandfonline.com/doi/full/10.1080/09515089.2024.2360673#abstract

McCormack, T., Burns, P., O'Connor, P., Jaroslawska, A., & Caruso, E. M. (2019). Do children and adolescents have a future-oriented bias? A developmental study of spontaneous and cued past and future thinking. *Psychological research*, *83*(4), 774–787. https://doi.org/10.1007/s00426-018-1077-5 ; https://pubmed.ncbi.nlm.nih.gov/30159672/#:~:text=Abstract,bias%20may%20be%20task%2Dspecific

**33**

WHO (2025). Adolescent Health. *World Health Organization*. https://www.who.int/southeastasia/health-topics/adolescent-health#:~:text=WHO%20defines%20'Adolescents'%20as%20individuals%20in%20the,countries%20of%20the%20South%2DEast%20Asia%20Region%20(SEAR).

Sawyer, S. M., Azzopardi, P. S., Wickremarathne, D., & Patton, G. C. (2018). The age of adolescence. *The Lancet. Child & adolescent health*, *2*(3), 223–228. https://doi.org/10.1016/S2352-4642(18)30022-1 ; https://pubmed.ncbi.nlm.nih.gov/30169257/#:~:text=Rather%20than%20age%2010%2D19,Psychology%2C%20Adolescent

Amy Peykoff Hardin, MD and Jesse M. Hackell, MD (2023). Age Limit of Pediatrics. *American Academy of Pediatrics.* PEDIATRICS Volume 140, number 3, September 2017:e20172151 From the American Academy of Pediatrics Organizational Principles to Guide and Define the Child Health Care System and/or Improve the Health of all Children POLICY STATEMENT This Policy Statement was reaffirmed

April 2023. Downloaded from http://publications.aap.org/pediatrics/article-pdf/140/3/e20172151/1688334/peds_20172151.pdf

34

McCormack, T., Burns, P., O'Connor, P., Jaroslawska, A., & Caruso, E. M. (2019). Do children and adolescents have a future-oriented bias? A developmental study of spontaneous and cued past and future thinking. *Psychological research*, *83*(4), 774–787. https://doi.org/10.1007/s00426-018-1077-5 ; https://pmc.ncbi.nlm.nih.gov/articles/PMC6529372/

McCormack, T., & Hoerl, C. (2020). Children's future-oriented cognition. In J. B. Benson (Ed.), *Advances in child development and behavior* (pp. 215–253). Elsevier Academic Press. https://doi.org/10.1016/bs.acdb.2020.01.008 ; https://www.sciencedirect.com/science/article/pii/S0065240720300082?via%3Dihub

35

McCormack, T., Burns, P., O'Connor, P., Jaroslawska, A., & Caruso, E. M. (2019). Do children and adolescents have a future-oriented bias? A developmental study of spontaneous and cued past and future thinking. *Psychological research*, *83*(4), 774–787. https://doi.org/10.1007/s00426-018-1077-5 ; https://pmc.ncbi.nlm.nih.gov/articles/PMC6529372/

McCormack, T., & Hoerl, C. (2020). Children's future-oriented cognition. In J. B. Benson (Ed.), *Advances in child development and behavior* (pp. 215–253). Elsevier Academic Press. https://doi.org/10.1016/bs.acdb.2020.01.008 ; https://www.sciencedirect.com/science/article/pii/S0065240720300082?via%3Dihub

36

Young, L., Bechara, A., Tranel, D., Damasio, H., Hauser, M., & Damasio, A. (2010). Damage to ventromedial prefrontal cortex impairs judgment of harmful intent. *Neuron*, *65*(6), 845–851. https://doi.org/10.1016/j.neuron.2010.03.003 ; https://pmc.ncbi.nlm.nih.gov/articles/PMC3085837/#:~:text =Summary,harmful%20intent%20for%20moral%20judgment.

37

**Kramer, Stephanie (2019). U.S. has world's highest rate of children living in single-parent households. Pew Research Center.** https://www.pewresearch.org/short-reads/2019/12/12/u-s-children-more-likely-than-children-in-other-countries-to-live-with-just-one-parent/

38

Chavda K, Nisarga V. Single Parenting: Impact on Child's Development. *Journal of Indian Association for Child and Adolescent Mental Health*. 2023;19(1):14-20. doi:10.1177/09731342231179017 ; https://journals.sagepub.com/doi/full/10.1177/09731342231 179017#:~:text=Various%20studies%20have%20found%20red uction,a%20positive%20attitude%20towards%20school.

Daryanani, I., Hamilton, J. L., Abramson, L. Y., & Alloy, L. B. (2016). Single Mother Parenting and Adolescent Psychopathology. *Journal of abnormal child psychology*, *44*(7), 1411–1423. https://doi.org/10.1007/s10802-016-0128-x ; https://pmc.ncbi.nlm.nih.gov/articles/PMC5226056/#:~:text =Across%20numerous%20studies%2C%20children%20raised, et%20al.%2C%202014).

39

United States Department of Justice. CRM 1-499. Juvenile Crime Facts. https://www.justice.gov/archives/jm/criminal-resource-manual-102-juvenile-crime-facts#:~:text=In%20all%2C%20twenty%2Dfive%20percent,Id.

40

Mendel, R. (2023). **Why Youth Incarceration Fails: An Updated Review of the Evidence.** The Sentencing Project.
https://www.sentencingproject.org/reports/why-youth-incarceration-fails-an-updated-review-of-the-evidence/

The Annie E. Casey Fountation (2019). What Are Status Offenses and Why Do They Matter?
https://www.aecf.org/blog/what-are-status-offenses-and-why-do-they-matter#:~:text=Most%20youths%20who%20engage%20in,The%20Annie%20E.

41

Lantz, Brendan and Knapp, Kyle G. (2024). Trends in Juvenile Offending
What You Need to Know. Council on Criminal Justice.
https://counciloncj.org/trends-in-juvenile-offending-what-you-need-to-know/

42

Lantz, Brendan and Knapp, Kyle G. (2024). Trends in Juvenile Offending
What You Need to Know. Council on Criminal Justice.
https://counciloncj.org/trends-in-juvenile-offending-what-you-need-to-know/

43

U.S. Census Bureau, 2009-2011 American Community Surveys, 2012 Condition of Children in Orange County, America's Families and Living Arrangements: 2012 by Jonathan Vespa and Jamie M. Lewis.    Direct Quote

44

U.S. Census Bureau, 2009-2011 American Community Surveys, 2012 Condition of Children in Orange County, America's Families and Living Arrangements: 2012 by Jonathan Vespa and Jamie M. Lewis. Direct Quote

45

Annie Casey Foundation (2024). Child Well-Being in Single-Parent Families. https://www.aecf.org/blog/child-well-being-in-single-parent-families#:~:text=While%20most%20children%20in%20single,adverse%20childhood%20experiences%20(ACEs). Direct Quote

46

U.S. Census Bureau, 2009-2011 American Community Surveys, 2012 Condition of Children in Orange County, America's Families and Living Arrangements: 2012 by Jonathan Vespa and Jamie M. Lewis.    Direct Quote

47

U.S. Census Bureau, 2009-2011 American Community Surveys, 2012 Condition of Children in Orange County, America's Families and Living Arrangements: 2012 by Jonathan Vespa and Jamie M. Lewis.    Direct Quote

48

Murray, C. (2022). These Are The Richest Americans Who Never Went To College. Forbes. https://www.forbes.com/sites/conormurray/2022/10/15/these-are-the-richest-americans-who-never-went-to-college/

49

McCormack, T., Burns, P., O'Connor, P. *et al.* Do children and adolescents have a future-oriented bias? A developmental study of spontaneous and cued past and future thinking. *Psychological Research* 83, 774–787 (2019). https://doi.org/10.1007/s00426-018-1077-5

50

Rizzo, A., Chaoyun, L. How young adults imagine their future? The role of temperamental traits. *Eur J Futures Res* **5,** 9 (2017). https://doi.org/10.1007/s40309-017-0116-6. Retrieved from:

https://eujournalfuturesresearch.springeropen.com/articles/10.1007/s40309-017-0116-6

51

Pelissolo, A., & Corruble, E. (2002). Personality factors in depressive disorders: contribution of the psychobiologic model developed by Cloninger. *L'encephale, 28*(4), 363-373.

Chen, C. Y., Lin, S. H., Li, P., Huang, W. L., & Lin, Y. H. (2015). The role of the harm avoidance personality in depression and anxiety during the medical internship. *Medicine, 94*(2), e389. https://doi.org/10.1097/MD.0000000000000389

Bajraktarov, S., Gudeva-Nikovska, D., Manuševa, N., & Arsova, S. (2017). Personality Characteristics as Predictive Factors for the Occurrence of Depressive Disorder. *Open access Macedonian journal of medical sciences, 5*(1), 48–53. https://doi.org/10.3889/oamjms.2017.022

52

Oettingen, G., & Mayer, D. (2002). The motivating function of thinking about the future: Expectations versus fantasies. *Journal of Personality and Social Psychology, 83*(5), 1198–1212. https://doi.org/10.1037/0022-3514.83.5.1198

53

Michelle L. Moulds, Eva Kandris, Susannah Starr, Amanda C.M. Wong (2007).
The relationship between rumination, avoidance and depression in a non-clinical sample,
*Behaviour Research and Therapy, Volume 45, Issue 2, 2007*, Pages 251-261, ISSN 0005-7967,
https://doi.org/10.1016/j.brat.2006.03.003 ; Retrieved from: https://www.sciencedirect.com/science/article/pii/S0005796706000611

Diana B. Henriques (2021). *Bernard Madoff, Architect of Largest Ponzi Scheme in History, Is Dead at 82*. The New York Times. Retrieved from:
https://www.nytimes.com/2021/04/14/business/bernie-madoff-dead.html

Biography (2021). Bernie Madoff. Biography. Retrieved from: https://www.biography.com/crime/bernard-madoff

Department of Justice (2022). Justice Department Announces Total Distribution of Over $4 Billion to Victims of Madoff Ponzi Scheme. Retrieved from:
https://www.justice.gov/opa/pr/justice-department-announces-total-distribution-over-4-billion-victims-madoff-ponzi-scheme

United States Attorney's Office (2020). United States V. Bernard L. Madoff And Related Cases. *United States Attorney's Office Southern District of New York*. Retrieved from:
https://www.justice.gov/usao-sdny/programs/victim-witness-services/united-states-v-bernard-l-madoff-and-related-cases

Taylor McNeil (2022). He Wrote the Book on Bernie Madoff. *Tufts University*. Retrieved from:
https://now.tufts.edu/2023/01/13/he-wrote-book-bernie-madoff

Netflix (2022). *Madoff: The Monster of Wall Street*, a popular new docuseries on Netflix. The series is based on the 2021 book *Madoff Talks: Uncovering the Untold Story Behind the Most Notorious Ponzi Scheme in History*, by Jim Campbell,

EMMA DOONEY (2023). How did Bernie Madoff get caught and when did the 'Monster of Wall Street' die?

*Woman & Home*. Retrieved from: https://www.womanandhome.com/life/news-entertainment/how-did-bernie-madoff-get-caught-and-when-did-the-monster-of-wall-street-die/

[56]

Raja Razek (2022). Bernie Madoff's sister and her husband dead in apparent murder-suicide. *CNN*. Retrieved from: https://www.cnn.com/2022/02/20/us/bernie-madoff-sister-husband-die/index.html#:~:text=Mark%20Madoff%2C%20the%20older%20son,his%20involvement%20in%20the%20scheme.

[57]

Markopolos, Harry (2010). *No One Would Listen: A True Financial Thriller*. Hoboken, New Jersey: John Wiley & Sons. ISBN 978-0-470-55373-2.

[58]

STELLA SECHOPOULOS (2022). Most in the U.S. say young adults today face more challenges than their parents' generation in some key areas. *Pew Research Center*. Retrieved from: https://www.pewresearch.org/fact-tank/2022/02/28/most-in-the-u-s-say-young-adults-today-face-more-challenges-than-their-parents-generation-in-some-key-areas/

[59]

"Evolution of wealth indicators, USA, 1913-2019". *WID.world. World Inequality Database. 2022. Archived from the original on July 5, 2023.*

Beeghley, Leonard (2004). The Structure of Social Stratification in the United States. Boston, MA: Allyn and Bacon. ISBN 0-205-37558-8.

Gilbert, Dennis (1998). *The American Class Structure*. New York: Wadsworth Publishing. ISBN 0-534-50520-1.

Thompson, William; Joseph Hickey (2005). *Society in Focus*. Boston, MA: Pearson. ISBN 0-205-41365-X.

60

APO. Marriage. *Aggieland Pregnancy Outreach*. Retrieved from: https://pregnancyoutreach.org/im-pregnant-2/marriage/#:~:text=In%20the%20past%2C%20many%20young,be%20divorced%20within%20six%20years.

61

Wang, Wendy (2020). The U.S. Divorce Rate Has Hit a 50-Year Low. *Institute for Family Studies*. Retrieved from: https://ifstudies.org/blog/the-us-divorce-rate-has-hit-a-50-year-low

62

Wang, Wendy (2020). The U.S. Divorce Rate Has Hit a 50-Year Low. *Institute for Family Studies*. Retrieved from: https://ifstudies.org/blog/the-us-divorce-rate-has-hit-a-50-year-low

63

Epps, David (2021). Divorce in America. *The Citizen*. Retrieved from: https://thecitizen.com/2021/11/03/divorce-in-america/#:~:text=In%201981%2C%20the%20divorce%20rate,for%202021%20will%20be%2045%25.

64

Elvira G. Aletta, Ph.D. (2021). What Makes a Family Functional vs Dysfunctional? *PsychCentral*. Retrieved from: https://psychcentral.com/blog/what-makes-a-family-functional-vs-dysfunctional#1

**65**

Herrera, P. M. (1997) La familia funcional y disfuncional, un indicador de salud. Rev Cubana Med Gen Integr; 13(6). Retrieved from: https://psychology-spot.com/functional-family-dysfunctional-family/

Delgado, Jennifer (2018).Functional Family vs. Dysfunctional family: 10 Characteristics that differentiate them. Psychology Spot. Retrieved from: https://psychology-spot.com/functional-family-dysfunctional-family/

**66**

Sowers, Pam (2002). Becoming parents: It's more than having a baby. UW News. Retrieved from: https://www.washington.edu/news/2002/06/05/becoming-parents-its-more-than-having-a-baby/

**67**

Olson, D. H. and Wilde, J. L. (Five Parenting Styles based on the Olson Circumplex Model. Retrieved from: https://app.prepare-enrich.com/pe/pdf/research/parenting_study.pdf

**68**

Brian D. Doss, Galena K. Rhoades, Scott M. Stanley, and Howard J. Markman (2009). The Effect of the Transition to Parenthood on Relationship Quality: An Eight-Year Prospective Study.
*J Pers Soc Psychol. 2009 Mar; 96(3): 601–619.* Retrieved from: https://www.ncbi.nlm.nih.gov/pmc/articles/PMC2702669/

**69**

Brian D. Doss, Galena K. Rhoades, Scott M. Stanley, and Howard J. Markman (2009). The Effect of the Transition to Parenthood on Relationship Quality: An Eight-Year Prospective Study.
*J Pers Soc Psychol. 2009 Mar; 96(3): 601–619.* Retrieved from: https://www.ncbi.nlm.nih.gov/pmc/articles/PMC2702669/

70

Kearney, Melissa. S. (2023). The elephant in the room. Brookings Institute. Retrieved from: https://www.brookings.edu/articles/the-elephant-in-the-room/?utm_campaign=Brookings%20Brief&utm_medium=email&utm_content=275161956&utm_source=hs_email

71

Kearney, Melissa. S. (2023). The elephant in the room. Brookings Institute. Retrieved from: https://www.brookings.edu/articles/the-elephant-in-the-room/?utm_campaign=Brookings%20Brief&utm_medium=email&utm_content=275161956&utm_source=hs_email

72

Wiebke Bleidorn, Christopher J. Hopwood, Mitja D. Back, Jaap J. A. Denissen, Marie Hennecke, Patrick L. Hill, Markus Jokela, Christian Kandler, Richard E. Lucas, Maike Luhmann, Ulrich Orth, Brent W. Roberts, Jenny Wagner, Cornelia Wrzus, Johannes Zimmermann (2021). Personality Trait Stability and Change. Retrieved from: https://ps.psychopen.eu/index.php/ps/article/view/6009/6009.html#:~:text=In%20summary%2C%20the%20literature%20on,and%20potentially%20also%20old%20age. ; https://ps.psychopen.eu/index.php/ps/article/view/6009/6009.html#:~:text=In%20summary%2C%20the%20literature%20on,and%20potentially%20also%20old%20age.

Bleidorn, W., Hopwood, C. J., Back, M. D., Denissen, J. J. A., Hennecke, M., Hill, P. L., Jokela, M., Kandler, C., Lucas, R. E., Luhmann, M., Orth, U., Roberts, B. W., Wagner, J., Wrzus, C., & Zimmermann, J. (2021). Personality Trait Stability and Change. *Personality Science, 2*, 1-20. https://doi.org/10.5964/ps.6009

Mathew A. Harris, Caroline E. Brett, Wendy Johnson and Ian J. Deary (2016). Personality Stability From Age 14 to Age 77 Years. Psychol Aging. 2016 Dec; 31(8): 862–874.
doi: 10.1037/pag0000133. PMCID: PMC5144810, PMID: 27929341. Retrieved from:
https://www.ncbi.nlm.nih.gov/pmc/articles/PMC5144810/#:~:text=Although%20lifelong%20personality%20stability%20has,across%20the%20entire%20life%20course.

73

Greenberg J., Pyszczynski T., Solomon S., Simon L., Breus M. Role of consciousness and accessibility death related thoughts in mortality salience effects. *J. Pers. Soc. Psychol.* 1994;67(4):627–637.

Greenberg J., Kosloff S., Solomon S., Cohen F., Landau M. Toward understanding the fame game: the effect of mortality salience on the apple of fame. *Self Ident.* 2010;9:1–8.

Hayes J., Schimel J., Faucher E.H., Williams T.J. Evidence for the DTA hypothesis II: threatening self-esteem increases death thought accessibility. *J. Exp. Soc. Psychol.* 2008;44:600–613.

74

Rybarski R, Bartczuk RP, Śliwak J, Zarzycka B. Religiosity and death anxiety among cancer patients: the mediating role of religious comfort and struggle. Int J Occup Med Environ Health. 2023 Nov 13;36(4):450-464. doi: 10.13075/ijomeh.1896.02094. Epub 2023 Sep 11. PMID: 37712529; PMCID: PMC10691416. Retrieved from:
https://www.ncbi.nlm.nih.gov/pmc/articles/PMC10691416/

Dadfar M and Lester D. Religiously, Spirituality and Death Anxiety. Austin J Psychiatry Behav Sci. 2017; 4(1): 1061. Retrieved from:
https://austinpublishinggroup.com/psychiatry-behavioral-sciences/fulltext/ajpbs-v4-id1061.php

David B. Feldman, Ian C. Fischer, Robert A. Gressis (2017). Does Religious Belief Matter for Grief and Death Anxiety? Experimental Philosophy Meets Psychology of Religion. Journal for the Scientific Study of Religion (55) 3, September 2016, Pages 531-539, First published: 27 January 2017. https://doi.org/10.1111/jssr.12288. Retrieved from: https://onlinelibrary.wiley.com/doi/full/10.1111/jssr.12288

Jonathan Jong,(2020). Chapter 2 - Death anxiety and religious belief: a critical review,
Editor(s): Kenneth E. Vail, Clay Routledge,. The Science of Religion, Spirituality, and Existentialism. Academic Press, 2020, Pages 21-35,, ISBN 9780128172049,
https://doi.org/10.1016/B978-0-12-817204-9.00003-2.
Retrieved from:
https://www.sciencedirect.com/science/article/pii/B9780128172049000032

75

Fitri, R. A., Asih, S. R., & Takwin, B. (2020). Social curiosity as a way to overcome death anxiety: perspective of terror management theory. *Heliyon, 6*(3), e03556.
https://doi.org/10.1016/j.heliyon.2020.e03556 . Retrieved from:
https://www.ncbi.nlm.nih.gov/pmc/articles/PMC7078517/

76

Gábor Király, Alexandra Köves (2023). Facing finitude: Death-awareness and sustainable transitions. Ecological Economics, Volume 205, 2023, 107729, ISSN 0921-8009.
https://doi.org/10.1016/j.ecolecon.2022.107729. Retrieved from:
https://www.sciencedirect.com/science/article/pii/S0921800922003901

77

Nencini, A., Meneghini, A. & Prati, M. (2015). Social Psychology and Performative Interventions in Human Systems. The GENERATIVE Method. Journal of Social Sciences, 11(1), 39-48. https://doi.org/10.3844/jssp.2015.39.48. Retrieved from:
https://thescipub.com/abstract/10.3844/jssp.2015.39.48

78

Nencini, A., Meneghini, A. & Prati, M. (2015). Social Psychology and Performative Interventions in Human Systems. The GENERATIVE Method. Journal of Social Sciences, 11(1), 39-48. https://doi.org/10.3844/jssp.2015.39.48. Retrieved from:
https://thescipub.com/abstract/10.3844/jssp.2015.39.48

79

Dijck, Jose van (2 January 2013). *The Culture of Connectivity: A Critical History of Social Media*. Oxford University Press. ISBN 978-0-19-997079-7.

80

Kietzmann, Jan H.; Hermkens, Kristopher (2011). *"Social media? Get serious! Understanding the functional building blocks of social media". Business Horizons (Submitted manuscript).* **54** *(3): 241– 251. doi:10.1016/j.bushor.2011.01.005. S2CID 51682132.*

Obar, Jonathan A.; Wildman, Steve (2015). *"Social media definition and the governance challenge: An introduction to the special issue". Telecommunications Policy.* **39** *(9): 745– 750. doi:10.2139/ssrn.2647377. ISSN 1556-5068. SSRN 2647377.*

*Kaplan, Andreas M.; Haenlein, Michael (2010). "Users of the world, unite! The challenges and opportunities of social media". Business Horizons.* **53** *(1). Bloomington, Indiana: Kelley School of Business: 61, 64–65,*

67. doi:*10.1016/j.bushor.2009.09.003*. *S2CID 16741539*.
*Retrieved 28 April 2019. Social Media is a very active and fast-moving domain. What may be up-to-date today could have disappeared from the virtual landscape tomorrow. It is therefore crucial for firms to have a set of guidelines that can be applied to any form of Social Media [...].*

Fuchs, Christian (2017). *Social media: a critical introduction (2nd ed.).* Los Angeles London New Delhi Singapore Washington DC Melbourne: SAGE. *ISBN 978-1-4739-6683-3.*

Boyd, Danah M.; Ellison, Nicole B. (2007). "Social Network Sites: Definition, History, and Scholarship". *Journal of Computer-Mediated Communication.* **13** *(1): 210–30. doi:10.1111/j.1083-6101.2007.00393.x*

81

Ashar, L. C. (2024). Social Media Impact: How Social Media Sites Affect Society.
*Business and Management Blog | American Public University*
https://www.apu.apus.edu/area-of-study/business-and-management/resources/how-social-media-sites-affect-society/

82

Ashar, L. C. (2024). Social Media Impact: How Social Media Sites Affect Society.
*Business and Management Blog | American Public University*
https://www.apu.apus.edu/area-of-study/business-and-management/resources/how-social-media-sites-affect-society/

83

Galvão, V.F., Maciel, C., Pereira, R. *et al.* Discussing human values in digital immortality: towards a value-oriented perspective. *J Braz Comput Soc* **27**, 15 (2021).

https://doi.org/10.1186/s13173-021-00121-x .
https://journal-bcs.springeropen.com/articles/10.1186/s13173-021-00121-x

[84]

Annie E. Casey Foundation Blog (2024). The Impact of social media and Technology on Gen Alpha. *The Annie E. Casey Foundation.* https://www.aecf.org/blog/impact-of-social-media-on-gen-alpha

[85]

Annie E. Casey Foundation Blog (2024). The Impact of social media and Technology on Gen Alpha. *The Annie E. Casey Foundation.* https://www.aecf.org/blog/impact-of-social-media-on-gen-alpha

[86]

Annie E. Casey Foundation Blog (2024). The Impact of social media and Technology on Gen Alpha. *The Annie E. Casey Foundation.* https://www.aecf.org/blog/impact-of-social-media-on-gen-alpha

[87]

Annie E. Casey Foundation Blog (2024). The Impact of social media and Technology on Gen Alpha. *The Annie E. Casey Foundation.* https://www.aecf.org/blog/impact-of-social-media-on-gen-alpha

[88]

Annie E. Casey Foundation Blog (2024). The Impact of social media and Technology on Gen Alpha. *The Annie E. Casey Foundation.* https://www.aecf.org/blog/impact-of-social-media-on-gen-alpha

[89]

Annie E. Casey Foundation Blog (2024). The Impact of social media and Technology on Gen Alpha. *The Annie E. Casey*

*Foundation.* https://www.aecf.org/blog/impact-of-social-media-on-gen-alpha

Annie E. Casey Foundation Blog (2024). The Impact of social media and Technology on Gen Alpha. *The Annie E. Casey Foundation.* https://www.aecf.org/blog/impact-of-social-media-on-gen-alpha

Ashar, L. C. (2024). Social Media Impact: How Social Media ites Affect Society. *American Public University.* https://www.apu.apus.edu/area-of-study/business-and-management/resources/how-social-media-sites-affect-society/

Canadian Department of Justice (2022). A Child's Age and Stage of Development Make a Difference. https://www.justice.gc.ca/eng/rp-pr/fl-lf/divorce/age/age2c.html#:~:text=Previous%20Page-,Pre%2Dteens%20(9%2D12%20years),friendships%20and%20extra%2Dcurricular%20activities

Piccerillo, L., Tescione, A., Iannaccone, A., & Digennaro, S. (2025). Alpha generation's social media use: sociocultural influences and emotional intelligence. International Journal of Adolescence and Youth, 30(1). https://doi.org/10.1080/02673843.2025.2454992. https://www.tandfonline.com/doi/full/10.1080/02673843.2025.2454992?af=R

Piccerillo, L., Tescione, A., Iannaccone, A., & Digennaro, S. (2025). Alpha generation's social media use: sociocultural influences and emotional intelligence. International Journal of Adolescence and Youth, 30(1).

https://doi.org/10.1080/02673843.2025.2454992.
https://www.tandfonline.com/doi/full/10.1080/02673843.2025.2454992?af=R

94

Statista (2024). Biggest social media platforms (2024). Statista. https://www.statista.com/statistics/272014/global-social-networks-ranked-by-number-of-users/

95

Statista (2024). Generation Alpha's social media usage according to their parents in the United States in 2024. https://www.statista.com/statistics/1546576/leading-social-media-platforms-gen-alpha-us/

96

Piccerillo, L., Tescione, A., Iannaccone, A., & Digennaro, S. (2025). Alpha generation's social media use: sociocultural influences and emotional intelligence. International Journal of Adolescence and Youth, 30(1).
https://doi.org/10.1080/02673843.2025.2454992.
https://www.tandfonline.com/doi/full/10.1080/02673843.2025.2454992?af=R

97

Turner, A. (2015). Generation Z: Technology and social interest. *Journal of Individual Psychology*, 71(2), 103–113.
https://doi.org/10.1353/jip.2015.0021 .
https://muse.jhu.edu/pub/15/article/586631/pdf .
https://muse.jhu.edu/article/586631

98

Social media usage can diminish social skills, particularly by reducing face-to-face interactions, which hinders the development of essential communication skills like reading nonverbal cues, building empathy, and managing real-time conversations. The asynchronous and curated nature of online interactions can also lead to social comparison, feelings of inadequacy, and a decrease in real-world social confidence. However, social media can also enhance social connectivity, provide community platforms, and support long-distance relationships, with the overall impact depending on mindful individual usage.

Bergman, M. (2025). How Does Social Media Affect Social Skills? *The Social Skills Center.* https://socialskillscenter.com/how-does-social-media-affect-social-skills/

How does social media affect teens' social skills? *Social Media Victims Law Center.* https://socialmediavictims.org/effects-of-social-media/teens-social-skills/#:~:text=Positive%20social%20media%20interactions%20can,in%20the%20%E2%80%9Creal%20world.%E2%80%9D

Kolhar, M., Kazi, R. N. A., & Alameen, A. (2021). Effect of social media use on learning, social interactions, and sleep duration among university students. *Saudi journal of biological sciences, 28*(4), 2216–2222. https://doi.org/10.1016/j.sjbs.2021.01.010 ; https://pmc.ncbi.nlm.nih.gov/articles/PMC8071811/

EBSCO (2024). Social Media Usage and Social Skills. https://www.ebsco.com/research-starters/social-sciences-and-humanities/social-media-usage-and-social-skills#full-article

Piccerillo, L., Tescione, A., Iannaccone, A., & Digennaro, S. (2025). Alpha generation's social media use: sociocultural influences and emotional intelligence. International Journal of Adolescence and Youth, 30(1). https://doi.org/10.1080/02673843.2025.2454992.

https://www.tandfonline.com/doi/full/10.1080/02673843.2025.2454992?af=R

100

May, R., (2024). Gen Alpha Social Media: What Publishers Need to Know.
Edited by Arabian, V. *State of Digital Publishing.*
https://www.stateofdigitalpublishing.com/audience-development/generation-alpha-and-social-media/

101

Vogels, Emily and Gelles-Watnick, Risa (2023). Teens and social media: Key findings from Pew Research Center surveys. Pew Research Center
https://www.pewresearch.org/short-reads/2023/04/24/teens-and-social-media-key-findings-from-pew-research-center-surveys/

102

Vogels, Emily and Gelles-Watnick, Risa (2023). Teens and social media: Key findings from Pew Research Center surveys. Pew Research Center
https://www.pewresearch.org/short-reads/2023/04/24/teens-and-social-media-key-findings-from-pew-research-center-surveys/

103

Vogels, Emily and Gelles-Watnick, Risa (2023). Teens and social media: Key findings from Pew Research Center surveys. Pew Research Center
https://www.pewresearch.org/short-reads/2023/04/24/teens-and-social-media-key-findings-from-pew-research-center-surveys/

104

Faverio, M., Anderson, M. and Park E. (2025). Teens, Social Media and Mental Health. Pew Research Center.

https://www.pewresearch.org/internet/2025/04/22/teens-social-media-and-mental-health/

105

Vogels, Emily and Gelles-Watnick, Risa (2023). Teens and social media: Key findings from Pew Research Center surveys. Pew Research Center
https://www.pewresearch.org/short-reads/2023/04/24/teens-and-social-media-key-findings-from-pew-research-center-surveys/

106

Vogels, Emily and Gelles-Watnick, Risa (2023). Teens and social media: Key findings from Pew Research Center surveys. Pew Research Center
https://www.pewresearch.org/short-reads/2023/04/24/teens-and-social-media-key-findings-from-pew-research-center-surveys/

107

Vogels, Emily and Gelles-Watnick, Risa (2023). Teens and social media: Key findings from Pew Research Center surveys. Pew Research Center
https://www.pewresearch.org/short-reads/2023/04/24/teens-and-social-media-key-findings-from-pew-research-center-surveys/

108

FBI Miami (2024). Sextortion: A Growing Threat Targeting Minors. https://www.fbi.gov/contact-us/field-offices/miami/news/sextortion-a-growing-threat-targeting-minors

109

Johnson, David (2024). Predators trying to connect with young teens using online platforms. How to protect your kids. https://www.wpxi.com/news/local/predators-trying-connect-with-young-teens-using-online-platforms-how-protect-your-kids/JTAXCLBMQJGOTKZ6NB3HT4HOSM/

Conservative reasoning estimates that approximately 2,000 extra teen suicides occur annually in the US due to social media and smartphone cyberbullying and harassment.

Serino, Kenichi (2025). WATCH: Meta whistleblowers testify on child safety research before Senate Judiciary Committee. PBS. https://www.pbs.org/newshour/politics/watch-live-meta-whistleblowers-testify-on-child-safety-research-before-senate-judiciary-committee

Senate Committee (2021-2022). S.Hrg. 117-769 — PROTECTING KIDS ONLINE: TESTIMONY FROM A FACEBOOK WHISTLEBLOWER. *S.Hrg. 117-769* . https://www.congress.gov/event/117th-congress/senate-event/330603/text

Facebook Whistleblower Frances Haugen Testifies. https://www.rev.com/transcripts/facebook-whistleblower-frances-haugen-testifies-on-children-social-media-use-full-senate-hearing-transcript

BARBARA ORTUTAY AND HALELUYA HADERO (2024). Meta. TikTok and other social media CEOs testify in heated Senate hearing on child exploitation. Associated Press. https://apnews.com/article/meta-tiktok-snap-discord-zuckerberg-testify-senate-00754a6bea92aaad62585ed55f219932

Softky, William (2024). Social Media Has a Colossal, Horrific Body Count. Fair Observer. https://www.fairobserver.com/world-news/social-media-has-a-colossal-horrific-body-count/

112

CDC (2024). 2023 Youth Risk Behavior Survey Results. *Centers for Disease Control.* https://www.cdc.gov/yrbs/results/2023-yrbs-results.html

113

Vermont Department of Health (2023). Youth Risk Behavior Survey (YRBS). https://www.healthvermont.gov/stats/population-health-surveys-data/youth-risk-behavior-survey-yrbs

114

Bergman, Matthew Attorney (2025). Social Media & Suicide. *Social Media Victims Law Center.* https://socialmediavictims.org/mental-health/suicide/

FastStats. (2023, January 18). FastStats – Suicide and Self-Harm Injury. https://www.cdc.gov/nchs/fastats/suicide.htm

Curtin, S. (2020, September 11). State Suicide Rates Among Adolescents and Young Adults Aged 10–24: United States. Centers for Disease Control and Prevention, 69.https://www.cdc.gov/nchs/data/nvsr/nvsr69/nvsr-69-11-508.pdf

115

C. (2017, August 3). QuickStats: Suicide Rates for Teens Aged 15–19 Years, by Sex — United. Centers for Disease Control and Prevention. https://www.cdc.gov/mmwr/volumes/66/wr/mm6630a6.htm

116

CDC (2025). *"Facts About Suicide".* www.cdc.gov. 2022-10-24. https://www.cdc.gov/suicide/facts/index.html

Niederkrotenthaler, Thomas; Stack, Steven, eds. (2017). Media and Suicide. doi:10.4324/9781351295246. ISBN 978-1-351-29523-9.

117

Softky, William (2024). Social Media Has a Colossal, Horrific Body Count. Fair Observer.
https://www.fairobserver.com/world-news/social-media-has-a-colossal-horrific-body-count/

118

Luxton, D. D., June, J. D., & Fairall, J. M. (2012). Social media and suicide: a public health perspective. *American journal of public health*, *102 Suppl 2*(Suppl 2), S195–S200.
https://doi.org/10.2105/AJPH.2011.300608 .
https://pmc.ncbi.nlm.nih.gov/articles/PMC3477910/

119

Luxton, D. D., June, J. D., & Fairall, J. M. (2012). Social media and suicide: a public health perspective. *American journal of public health*, *102 Suppl 2*(Suppl 2), S195–S200.
https://doi.org/10.2105/AJPH.2011.300608.
https://pmc.ncbi.nlm.nih.gov/articles/PMC3477910/

120

Katella Kathy (2024). How Social Media Affects Your Teen's Mental Health: A Parent's Guide. Yale Medicine.
https://www.yalemedicine.org/news/social-media-teen-mental-health-a-parents-guide

121

Office of the Surgeon General (2023). Social Media and Youth Mental Health. United States Office of the Surgeon General.
https://www.hhs.gov/surgeongeneral/reports-and-publications/youth-mental-health/social-media/index.html .
https://www.hhs.gov/sites/default/files/sg-youth-mental-health-social-media-summary.pdf

122

Dunlop, S. M., More, E., & Romer, D. (2011). Where do youth learn about suicides on the Internet, and what influence does this have on suicidal ideation?. *Journal of child psychology and psychiatry, and allied disciplines, 52*(10), 1073–1080. https://doi.org/10.1111/j.1469-7610.2011.02416.x . https://pubmed.ncbi.nlm.nih.gov/21658185/

123

Luxton, D. D., June, J. D., & Fairall, J. M. (2012). Social media and suicide: a public health perspective. *American journal of public health, 102 Suppl 2*(Suppl 2), S195–S200. https://doi.org/10.2105/AJPH.2011.300608. https://pmc.ncbi.nlm.nih.gov/articles/PMC3477910/

124

Luxton, D. D., June, J. D., & Fairall, J. M. (2012). Social media and suicide: a public health perspective. *American journal of public health, 102 Suppl 2*(Suppl 2), S195–S200. https://doi.org/10.2105/AJPH.2011.300608. https://pmc.ncbi.nlm.nih.gov/articles/PMC3477910/

125

Pappas S. (2023). More than 20% of teens have seriously considered suicide. Psychologists and communities can help tackle the problem. American Psychological Association, Vol. 5, No. 5, P. 54. https://www.apa.org/monitor/2023/07/psychologists-preventing-teen-suicide

126

Memon, A. M., Sharma, S. G., Mohite, S. S., & Jain, S. (2018). The role of online social networking on deliberate self-harm and suicidality in adolescents: A systematized review of literature. *Indian journal of psychiatry, 60*(4), 384–392. https://doi.org/10.4103/psychiatry.IndianJPsychiatry_414_17 .

https://pmc.ncbi.nlm.nih.gov/articles/PMC6278213/#:~:text=Abstract,engage%20in%20self%2Dharm%20behavior.

Luxton, D. D., June, J. D., & Fairall, J. M. (2012). Social media and suicide: a public health perspective. *American journal of public health*, *102 Suppl 2*(Suppl 2), S195–S200.
https://doi.org/10.2105/AJPH.2011.300608
https://pmc.ncbi.nlm.nih.gov/articles/PMC3477910/#:~:text=Suicide%20is%20a%20considerable%20public,national%20attention%20to%20this%20topic.

Bergman, Matthew Attorney (2025). Social Media & Suicide. *Social Media Victims Law Center*.
https://socialmediavictims.org/mental-health/suicide/

Memon, A. M., Sharma, S. G., Mohite, S. S., & Jain, S. (2018). The role of online social networking on deliberate self-harm and suicidality in adolescents: A systematized review of literature. *Indian journal of psychiatry*, *60*(4), 384–392.
https://doi.org/10.4103/psychiatry.IndianJPsychiatry_414_17

https://pmc.ncbi.nlm.nih.gov/articles/PMC6278213/#:~:text=Abstract,engage%20in%20self%2Dharm%20behavior.

Luxton, D. D., June, J. D., & Fairall, J. M. (2012). Social media and suicide: a public health perspective. *American journal of public health*, *102 Suppl 2*(Suppl 2), S195–S200.
https://doi.org/10.2105/AJPH.2011.300608
https://pmc.ncbi.nlm.nih.gov/articles/PMC3477910/#:~:text=Suicide%20is%20a%20considerable%20public,national%20attention%20to%20this%20topic.

Bergman, Matthew Attorney (2025). Social Media & Suicide. *Social Media Victims Law Center*.
https://socialmediavictims.org/mental-health/suicide/

Martin, James (2025). 18 Cyberbullying Facts & Statistics (2025). https://explodingtopics.com/blog/cyberbullying-stats

128

Memon, A. M., Sharma, S. G., Mohite, S. S., & Jain, S. (2018). The role of online social networking on deliberate self-harm and suicidality in adolescents: A systematized review of literature. *Indian journal of psychiatry, 60*(4), 384–392. https://doi.org/10.4103/psychiatry.IndianJPsychiatry_414_17
.
https://pmc.ncbi.nlm.nih.gov/articles/PMC6278213/#:~:text=Abstract,engage%20in%20self%2Dharm%20behavior.

Luxton, D. D., June, J. D., & Fairall, J. M. (2012). Social media and suicide: a public health perspective. *American journal of public health, 102 Suppl 2*(Suppl 2), S195–S200. https://doi.org/10.2105/AJPH.2011.300608
https://pmc.ncbi.nlm.nih.gov/articles/PMC3477910/#:~:text=Suicide%20is%20a%20considerable%20public,national%20attention%20to%20this%20topic.

Bergman, Matthew Attorney (2025). Social Media & Suicide. *Social Media Victims Law Center.*
https://socialmediavictims.org/mental-health/suicide/

Martin, James (2025). 18 Cyberbullying Facts & Statistics (2025). https://explodingtopics.com/blog/cyberbullying-stats

129

Memon, A. M., Sharma, S. G., Mohite, S. S., & Jain, S. (2018). The role of online social networking on deliberate self-harm and suicidality in adolescents: A systematized review of literature. *Indian journal of psychiatry, 60*(4), 384–392. https://doi.org/10.4103/psychiatry.IndianJPsychiatry_414_17
.
https://pmc.ncbi.nlm.nih.gov/articles/PMC6278213/#:~:text=Abstract,engage%20in%20self%2Dharm%20behavior.

Luxton, D. D., June, J. D., & Fairall, J. M. (2012). Social media and suicide: a public health perspective. *American journal of public health*, *102 Suppl 2*(Suppl 2), S195–S200.
https://doi.org/10.2105/AJPH.2011.300608
https://pmc.ncbi.nlm.nih.gov/articles/PMC3477910/#:~:text=Suicide%20is%20a%20considerable%20public,national%20attention%20to%20this%20topic.

Bergman, Matthew Attorney (2025). Social Media & Suicide. *Social Media Victims Law Center*.
https://socialmediavictims.org/mental-health/suicide/

Martin, James (2025). 18 Cyberbullying Facts & Statistics (2025). https://explodingtopics.com/blog/cyberbullying-stats

Memon, A. M., Sharma, S. G., Mohite, S. S., & Jain, S. (2018). The role of online social networking on deliberate self-harm and suicidality in adolescents: A systematized review of literature. *Indian journal of psychiatry*, *60*(4), 384–392.
https://doi.org/10.4103/psychiatry.IndianJPsychiatry_414_17
https://pmc.ncbi.nlm.nih.gov/articles/PMC6278213/#:~:text=Abstract,engage%20in%20self%2Dharm%20behavior.

Luxton, D. D., June, J. D., & Fairall, J. M. (2012). Social media and suicide: a public health perspective. *American journal of public health*, *102 Suppl 2*(Suppl 2), S195–S200.
https://doi.org/10.2105/AJPH.2011.300608
https://pmc.ncbi.nlm.nih.gov/articles/PMC3477910/#:~:text=Suicide%20is%20a%20considerable%20public,national%20attention%20to%20this%20topic.

Bergman, Matthew Attorney (2025). Social Media & Suicide. *Social Media Victims Law Center*.
https://socialmediavictims.org/mental-health/suicide/

Martin, James (2025). 18 Cyberbullying Facts & Statistics (2025). https://explodingtopics.com/blog/cyberbullying-stats

131

Jocelyn I. Meza, Ph.D., Katie Patel, M.S., Eraka Bath, M.D (2022). Black Youth Suicide Crisis: Prevalence Rates,Review of Risk and Protective Factors, and Current Evidence-Based Practices, FOCUS, Volume 20, Number 2, https://doi.org/10.1176/appi.focus.20210034 . https://psychiatryonline.org/doi/10.1176/appi.focus.20210034 . https://psychiatryonline.org/doi/epdf/10.1176/appi.focus.20210034

132

Prinstein, M. J., Nesi, J., & Telzer, E. H. (2020). Commentary: An updated agenda for the study of digital media use and adolescent development—Future directions following Odgers & Jensen (2020). *Journal of Child Psychology and Psychiatry, 61*(3), 349–352. https://doi.org/10.1111/jcpp.13219

133

Magis-Weinberg, L., Ballonoff Suleiman, A., & Dahl, R. E. (2021). Context, development, and digital media: Implications for very young adolescents in LMICs. *Frontiers in Psychology, 12*, Article 632713. https://doi.org/10.3389/fpsyg.2021.632713 https://academic.oup.com/joc/article/63/2/221/4035964?login=true

Orben, A., Przybylski, A. K., Blakemore, S.-J., Kievit, R. A. (2022). Windows of developmental sensitivity to social media. *Nature Communications, 13*(1649). https://doi.org/10.1038/s41467-022-29296-3 . https://www.nature.com/articles/s41467-022-29296-3

Orben, A., & Blakemore, S.-J. (2023). How social media affects teen mental health: A missing link. *Nature, 614*(7948), 410–412. https://doi.org/10.1038/d41586-023-00402-9 . https://www.nature.com/articles/d41586-023-00402-9

Somerville, L. H., & Casey, B. J. (2010). Developmental neurobiology of cognitive control and motivational systems. *Current Opinion in Neurobiology, 20*(2), 236–241. https://doi.org/10.1016/j.conb.2010.01.006 . https://www.sciencedirect.com/science/article/pii/S0959438810000073?via%3Dihub

Noble, S. U. (2018). *Algorithms of oppression: How search engines reinforce racism.* New York University Press. https://doi.org/10.18574/nyu/9781479833641.001.0001 . https://www.degruyterbrill.com/document/doi/10.18574/nyu/9781479833641.001.0001/html

Relia, K., Li, Z., Cook, S. H., & Chunara, R. (2019). Race, ethnicity and national origin-based discrimination in social media and hate crimes across 100 U.S. cities. *Proceedings of the International AAAI Conference on Web and Social Media, 13*, 417–427. https://doi.org/10.1609/icwsm.v13i01.3354 . https://ojs.aaai.org/index.php/ICWSM/article/view/3354

Tynes, B. M., Willis, H. A., Stewart, A. M., & Hamilton, M. W. (2019). Race-related traumatic events online and mental health among adolescents of color. *The Journal of Adolescent Health, 65*(3), 371–377. https://doi.org/10.1016/j.jadohealth.2019.03.006 . https://www.sciencedirect.com/science/article/pii/S1054139X19301648?via%3Dihub

Charmaraman, L., Lynch, A. D., Richer, A. M., & Zhai, E. (2022). Examining early adolescent positive and negative social technology behaviors and well-being during the COVID-19 pandemic. *Technology, Mind, and Behavior, 3*(1). https://doi org/10.1037/tmb0000062 . https://tmb.apaopen.org/pub/x0dln2pf/release/2

Magis-Weinberg, L., Gys, C. L., Berger, E. L., Domoff, S. E., & Dahl, R. E. (2021). Positive and negative online experiences and loneliness in Peruvian adolescents during the COVID-19 lockdown. *Journal of Research on Adolescence, 31*(3), 717–733. https://doi.org/10.1111/jora.12666 . https://onlinelibrary.wiley.com/doi/10.1111/jora.12666

Psihogios, A. M., Ahmed, A. M., McKelvey, E. R., Toto, D., Avila, I., Hekimian-Brogan, E., Steward, Z., Schwartz, L. A., & Barakat, L. P. (2022). Social media to promote treatment adherence among adolescents and young adults with chronic health conditions: A topical review and TikTok application. *Clinical Practice in Pediatric Psychology, 17*(4), 440–451. https://doi.org/10.1037/cpp0000459 . https://journals.sagepub.com/doi/10.1037/cpp0000459

Holtz, B. E., & Kanthawala, S. (2020). #T1DLooksLikeMe: Exploring self-disclosure, social support, and Type 1 diabetes on Instagram. *Frontiers in Communication, 5.* https://doi.org/10.3389/fcomm.2020.510278 . https://www.frontiersin.org/journals/communication/articles/10.3389/fcomm.2020.510278/full

Nesi, J., Mann, S. and Robb, M. B. (2023). Teens and mental health: How girls really feel about social media. San Francisco, California: Common Sense

Armstrong-Carter, E., & Telzer, E. H. (2021). Advancing measurement and research on youths' prosocial behavior in

the digital age. *Child Development Perspectives*, *15*(1), 31–36. https://doi.org/10.1111/cdep.12396 .
https://srcd.onlinelibrary.wiley.com/doi/10.1111/cdep.12396

Massing-Schaffer, M., Nesi, J., Telzer, E. H., Lindquist, K. A., & Prinstein, M. J. (2022). Adolescent peer experiences and prospective suicidal ideation: The protective role of online-only friendships. *Journal of Clinical Child and Adolescent Psychology*, *51*(1), 49–60. https://doi.org/10.1080/15374416.2020.1750019 .
https://www.tandfonline.com/doi/full/10.1080/15374416.2020.1750019

Craig, S. L., Eaton, A. D., McInroy, L. B., Leung, V. W. Y., & Krishnan, S. (2021). Can social media participation enhance LGBTQ+ youth well-being? Development of the Social Media Benefits Scale. *Social Media + Society*, *7*(1), 205630512198893. https://doi.org/10.1177/2056305121988931 .
https://journals.sagepub.com/doi/10.1177/2056305121988931

APA (2023). Health Advisory on Social Media Use in Adolescence. American Psychological Association.
https://www.apa.org/topics/social-media-internet/health-advisory-adolescent-social-media-use.pdf .
https://www.apa.org/topics/social-media-internet/health-advisory-adolescent-social-media-use

APA (2024). Potential Risks of Content, Features, and Functions. American Psychological Association.
https://www.apa.org/topics/social-media-internet/psychological-science-behind-youth-social-media.pdf .
https://www.apa.org/topics/social-media-internet/youth-social-media-2024

Champion, C. (2023). Is social media causing psychological harm to youth and young adults? UCLA Health.
https://www.uclahealth.org/news/article/social-media-causing-psychological-harm-youth-and-young

Miller, C., Bubrick, J. Hamlet, A. (2025). Does Social Media Use Cause Depression? Child Mind Institute.
https://childmind.org/article/is-social-media-use-causing-depression/

[138]

Price, C. 2025). 25 Alternative Search Engines You Can Use Instead Of Google.
https://www.searchenginejournal.com/alternative-search-engines/271409/#youcom

[139]

Ku, Daniel (2020). Generation Alpha: How Will They Use The Internet? *Search Engine Journal.*
https://www.postbeyond.com/blog/generation-alpha-predictions/

[140]

"We live in a society where viewpoints are more important than Human Decency"
https://www.tiktok.com/@aubrieina/video/7505980638546119966

[141]

Pappas S. (2023). More than 20% of teens have seriously considered suicide. Psychologists and communities can help tackle the problem. American Psychological Association, Vol. 5, No. 5, P. 54.
https://www.apa.org/monitor/2023/07/psychologists-preventing-teen-suicide

Dr. Ronald Barnes

www.ingramcontent.com/pod-product-compliance
Lightning Source LLC
LaVergne TN
LVHW020133080526
838202LV00047B/3930